Chemistry and Technology of Biodegradable Polymers

Chemistry and Technology of Biodegradable Polymers

Edited by

G. J. L. GRIFFIN
Director
Epron Industries Ltd
Stamford

BLACKIE ACADEMIC & PROFESSIONAL
An Imprint of Chapman & Hall

London · Glasgow · New York · Tokyo · Melbourne · Madras

Published by
Blackie Academic and Professional, an imprint of Chapman & Hall,
Wester Cleddens Road, Bishopbriggs, Glasgow G64 2NZ

Chapman & Hall, 2–6 Boundary Row, London SE1 8HN, UK

Blackie Academic & Professional, Wester Cleddens Road, Bishopbriggs, Glasgow G64 2NZ, UK

Chapman & Hall Inc., One Penn Plaza, 41st Floor, New York NY10119, USA

Chapman & Hall Japan, Thomson Publishing Japan, Hirakawacho Nemoto Building, 6F, 1–7–11 Hirakawa-cho, Chiyoda-ku, Tokyo 102, Japan

DA Book (Aust.) Pty Ltd, 648 Whitehorse Road, Mitcham 3132, Victoria, Australia

Chapman & Hall India, R. Seshadri, 32 Second Main Road, CIT East, Madras 600 035, India

First edition 1994

© 1994 Chapman & Hall

Phototypeset in 10/12pt Times by Intype, London
Printed in Great Britain by St Edmundsbury Press, Bury St. Edmunds, Suffolk

ISBN 0 7514 0003 3

A catalogue record for this book is available from the British Library.

Library of Congress Cataloging-in-Publication data

Chemistry and technology of biodegradable polymers / edited by G. J. L. Griffin. — 1st ed.
   p.   cm.
  Includes bibliographical references and index.
  ISBN 0–7514–0003–3
  1. Polymers—Biodegradation.   I. Griffin, G. J. L.
QP801.P64C48 1994
620.1′9204223—dc20   93–33974
                                        CIP

∞ Printed on acid-free text paper, manufactured in accordance with ANSI/ NISO Z39.48–1992 (Permanence of Paper)

Preface

Since the early 1970s the subject of biodegradable plastics has acquired a rapidly growing literature of academic research papers. It has also acquired a formidable volume of patent documentation and all this has been over-whelmed by an astonishing quantity of serious media and political comment. A new entrant into any technical arena would, in most technologies, simply visit their technical library and pick up a text book on the subject in the expectation of absorbing the basic facts before launching into the daily task of updating and evaluating. Scientific conferences have produced many substantial volumes carrying the word 'biodegradable' on their covers, and there has even been a specialist monograph on the topic of bacterially produced polymers but, surprisingly, no book has yet emerged providing a general survey of the subject. Having devoted half my professional career to the subject of biodegradable plastics I agreed to take on the editorial job of producing such a book when asked by the publisher. I knew that the task of finding expert specialists and persuading them to contribute dispassionate accounts of their specialisms would not be easy, but the difficulties that I have encountered were far greater than I expected. Some were simply too busy, others were involved in patent disputes or commercial negotiations. In giving an account of the work that I and my students carried out at Brunel University I believe that I have written in a manner that displays enthusiasm without prejudice. I have tried to treat the enthusiasms of those that built up the work on gelatinised starch blends in the same manner and I must record my gratitude to Dr Catia Bastioli of Novamont spa, to Dr William Doane and his colleagues of the USDA, and to Dr Andrew Hopkins of Warner Lambert Novon division for guiding my hand and kindly providing all the hard data that I insisted on including. Many patents are referred to throughout this book because they are important sources of technical information often not obtainable from the academic journals. Experienced readers will know how to study these curious documents with their mixtures of technology and legalities and no attempt is made in these pages to judge the strength and validity of the patents associated with the various topics.

A plastic bag for peanuts does not really need to last for a century and the environmental campaigners had little trouble in convincing the plastics technologists that properties of the products of their skills needed adjustment. Some of the same campaigners have reacted against the consequences of their own success when they realised that any new technology must come to the market through the same route as the original plastics packs and a few companies have used sales promotion techniques that have created some obvious targets for criticism. Sadly the consequence of this reaction has been to diminish the flow of funds to support the very research that was originally demanded. We have since been able to watch the parallel business of recycling plastics go through the same evolution, from being a normal part of every plastic technologist's professional knowledge to becoming a media and political obsession with embarrassing consequences in certain parts of Europe. I have to express special gratitude to Charles Armistead of Manchester Plastics (USA) for agreeing to write an account of how he has been recycling degradable plastics since he started manufacturing the products five years ago. Green enthusiasts have declared the recycling of degradable plastics as technically impossible. I feel sure that this opinion of what he was doing as daily routine almost certainly influenced Mr Armistead towards contributing to this book. Being myself only too well aware of the pressures of work on factory managers makes me even more appreciative of his collaboration.

One gratifying aspect of the present state of the biodegradable plastics art is that the business of testing and standardisation of the products is emerging step by step into a formal body of knowledge. This development made it possible for me to invite Dr Seal of Euro Laboratories (UK) to write the chapter on testing, knowing that he has made the subject his specialism over many years and I thank him for taking time from his heavy schedule as chief scientist and administrator of his microbiology laboratories.

Serious environmentalists have very properly pointed out the importance of establishing that new products which find their way into the waste stream should not give rise to new toxicological hazards. I was, therefore, delighted when I succeeded in persuading Professor Albertsson, of the Royal Technical Institute in Stockholm, to contribute her chapter on the chemistry and biochemistry of the degradation process and I am most grateful to her.

Guided by the original plan to collect together hard information on these degradable plastics which are currently taken seriously in the commercial arena the obvious need to include material on the bacterially generated thermoplastics had to be respected. My problem was that of finding an author whose enthusiasm for the academic fascination of these substances did not preclude an ability to assess them as commercial plastics. My gratitude to Professor Marchessault and his colleague Phillipa Hocking is considerable.

Some readers may be familiar with the mechanics of producing technical books and will know that the labour is hugely increased when several authors are involved, and the more so when they are scattered around the world, despite the availability of various digital electronic communication techniques. If all the authors were working in established university science faculties then the whole job could probably have been completed in one central computer. A technological topic, however, finds its enthusiasts too widely scattered for the computer approach to work and this book owes much to Federal Express and the fax machine. It owes even more to the editorial staff of Blackie Academic and Professional who have contributed great assistance in collating the text and dealing with the proofs against a fierce print deadline quite apart from advising and comforting this editor at moments when the whole project seemed impossible. I now view the final page of the finished work with relief but with the thought that I must now quickly scan all this month's technical journals and revise the whole book—a concern aggravated by the recent commercial launch of the Japanese biodegradable polyester material ˙Bionolle™ too late for inclusion. I am assured, however, that this is a common affliction of technical authors and I must hope that whatever weaknesses there may have been in my planning of the work, the finished product will be of some service to its young technology and the associated industry—it may even justify the future contemplation of a second edition.

G.J.L.G.

Contributors

Professor A-C. Albertsson	Department of Polymer Technology, Royal Institute of Technology, S–100 44 Stockholm, Sweden
Mr C. R. Armistead	Manchester Packaging Co., 2000 East James Boulevard, St James, MO 65559, USA
Mr G. J. L. Griffin	Ecological Materials Research Institute, Epron Industries Ltd, Units 4–6 Ketton Business Estate, Ketton, Stamford, Lincolnshire PE9 3SZ, UK
Ms P. J. Hocking	McGill University, Department of Chemistry, Pulp and Paper Building, 3420 University Street, Montreal, Quebec H3A 2A7, Canada
Dr S. Karlsson	Department of Polymer Technology, Royal Institute of Technology, S–100 44. Stockholm, Sweden
Professor R. H. Marchessault	McGill University, Department of Chemistry, Pulp and Paper Building, 3420 University Street, Montreal, Quebec H3A 2A7, Canada
Dr K. J. Seal	Euro Laboratories Ltd, 20 Howard Way, Newport Pagnell, Buckinghamshire MK16 9QS, UK

Contents

1 Introduction   **1**
G. J. L. GRIFFIN

2 Chemistry and biochemistry of polymer biodegradation   **7**
A-C. ALBERTSSON and S. KARLSSON

2.1 Introduction	7
2.2 Enzymes	8
2.2.1 Enzyme nomenclature	8
2.2.2 Enzyme specificity	9
2.2.3 Physical factors affecting the activity of enzymes	9
2.2.4 Enzyme mechanisms	9
2.3 Chemical degradation initiates biodegradation	11
2.3.1 The biodegradation of polyolefins	13
2.3.2 Hydrophobicity influences biodegradability	15
2.4 Hydrolysis of synthetic biodegradable polymers	15
References	17

3 Particulate starch based products   **18**
G. J. L. GRIFFIN

3.1 Development of the technology	18
3.2 Current objectives	23
3.3 Relevant starch technology	25
3.4 Manufacture of masterbatch	27
3.5 Conversion technology	30
3.5.1 Processing precautions—moisture	30
3.5.2 Processing precautions—temperature	31
3.5.3 Rheological considerations	31
3.5.4 Cyclic conversion processes	33
3.6 Physical properties of products	33
3.6.1 Sample preparation	34
3.6.2 Physical testing methods	34
3.6.3 Test results	35
3.7 Quality control testing of degradation	39

      3.7.1 Autoxidation measurement   40

3.7.1 Autoxidation measurement	40
3.7.2 Biodegradation assessment	44
3.7.3 Soil burial tests	45
References	46
4 Biopolyesters	**48**
P. J. HOCKING and R. H. MARCHESSAULT	
4.1 Introduction	48
4.2 History	49
4.3 Biosynthesis	51
4.4 Isolation	55
4.4.1 Solvent extraction	55
4.4.2 Sodium hypochlorite digestion	56
4.4.3 Enzymatic digestion	57
4.5 Properties	57
4.5.1 Crystal structure	61
4.5.2 Nascent morphology	65
4.6 Degradation	68
4.6.1 Intracellular biodegradation	68
4.6.2 Extracellular biodegradation	71
4.6.3 Thermal degradation	72
4.6.4 Hydrolytic degradation	72
4.6.5 Environmental degradation	73
4.6.6 Effects on recycling	75
4.7 Applications	75
4.8 Economics	78
4.9 Future prospects	80
4.9.1 Other bacterial PHAs	80
4.9.2 Synthetic PHAs	84
4.9.3 Blends	86
4.9.4 Genetic engineering	87
4.9.5 Commercial developments	88
Acknowledgements	89
References	89
5 Recycling technology for biodegradable plastics	**97**
C. R. ARMISTEAD	
5.1 Introduction	97
5.2 Conventional recycling	98
5.2.1 Economic incentive	98
5.2.2 Recycling problems	98
5.3 Degradables complicate recycling	99

| | | 5.3.1 Polyethylene/corn starch film | 99 |

5.4   Reprocessing polyethylene/corn starch film scrap   103
     5.4.1   Learning to reprocess PE/S   103
     5.4.2   Calcium oxide moisture scavenger   105
     5.4.3   Temperature control   105
     5.4.4   Accounting for pro-oxidant   105
     5.4.5   Handling PE/S repro   106
5.5   Economics of in-plant recycling   107
5.6   Using PE/S repro   108
     5.6.1   Comparative study of PE/S repro on film properties   108
     5.6.2   Does it make sense to recycle degradable PE/S?   111
     5.6.3   Recycling other degradables   112
Notes   113
References   115

6   Test methods and standards for biodegradable plastics   116
K. J. SEAL
6.1   Introduction   116
6.2   Defining biodegradability   117
6.3   Criteria used in the evaluation of biodegradable polymers   118
6.4   Tiered systems for evaluating biodegradability   119
6.5   Choice of environment   120
6.6   Choosing the most appropriate methodology   121
6.7   Description of current test methods   123
     6.7.1   Screening tests for ready biodegradability   123
     6.7.2   Tests for inherent biodegradability   126
     6.7.3   Tests for simulation studies   130
6.8   Other methods for assessing polymer biodegradability   130
     6.8.1   Petri dish screen   130
     6.8.2   Environmental chamber method   131
     6.8.3   Soil burial tests   132
6.9   Test method developments for the future   133
References   134

7   Gelatinised starch products   135
G. J. L. GRIFFIN
7.1   History   135
7.2   Gelatinised starch blends with synthetic polymers   137
     7.2.1   Mater-Bi™ film extrusion   139
     7.2.2   Mater-Bi™ injection moulding   141
     7.2.3   Properties of products   145
7.3   High starch content products   148
References   150

Index   151

1   Introduction

G. J. L. GRIFFIN

Polystyrene, polyethylene and polyvinyl chloride became technologically significant during World War II in special electronics applications, or in chemical engineering plant construction. Although small amounts of cellulose ester films were being used in retail packaging before that time, it is unlikely that anyone could have foreseen the way in which these highly specialised synthetic polymers would soon become commonplace materials enveloping almost every piece of foodstuff or domestic article that we buy. The great revolution in retail distributive methods over the past 40 years, triggered by the growth of a car owning urban population, and under pressure from public health authorities and greater consumer discrimination, has irreversibly created a demand for accurate automated and hygienic packaging techniques and appropriate packaging materials.

The plastics industry, fastest growing of the materials industries, responded by developing special grades of polymers and associated machine processes such as high speed thin wall injection moulding, extrusion blowing of thin films and the extrusion blowing of bottles. The plastics packaging activity became a recognised industry in its own right and was an impressive technical and commercial success. The only point overlooked was the ultimate fate of these millions of pots, packets and bottles made mostly of polyolefin materials. The comforting belief and assumption was that such items would spread out from the factories through wholesalers, retailers and retail customers until, diluted by some inverse power law, they finally disappeared. This thinking was rudely disturbed by a growing chorus of complaints from various quarters reporting that these plastic packs were focusing and concentrating in awkward and often unlikely places such as the Sargasso sea, or beauty spots on the Scottish islands. Shoals of pink plastic tampon applicators floated onto the beaches of New Jersey and even local authority landfill sites reported extra costs in clearing their boundary fences from windblown plastics items.

Public awareness of the uncomfortable fact that infinite dilution does not work as a means of 'losing' unwanted materials had already been alerted by the powerful impact of Rachel Carson's book *Silent Spring*. This was published in 1962 and went rapidly into paperback editions and

magazine reprints. It is true that her concern was about dangerous pesticides rather than the fate of inert and non-toxic plastics, but the element
of her message about a finite world came to have obvious implications in
many areas of activity of a society based increasingly upon disposable
items. Even larger personal possessions are not exempt: cars with an
expected lifetime of ten years and washing machines of about six are both
items with a growing content of plastics materials. The commentators of
the period make it clear that the giants of the petrochemical industry were
caught off guard by the growing protests. They announced that plastics
were non toxic and virtually indestructible, and therefore, they could cause
no mischief. Having only just succeeded, by the 1960s, in suppressing the
postwar view of plastics as 'ersatz' or inferior substitutes for timber or
natural textiles, this new battleground was distinctly unwelcome. A useful
survey of the situation of that period is to be found in the monograph
Disposal of plastics waste and litter by Staudinger published in 1970 by the
Society of Chemical Industry as part of a publication programme initiated
in 1964 by their president Lord Fleck — a programme concerned with the
impact of the chemical industry on the 'public at large'. Staudinger draws
attention specifically to his belief that 'synthetic polymers are not biodegradable', a conclusion drawn from the considerable literature of biodeterioration, and he makes no reference to the possible destruction of plastics
litter by photodegradation.

When attempts to identify microorganisms capable of digesting synthetic
polymers failed, some attention was paid to the possibilities of encouraging
mutants that might have this ability. Microbiologists and geneticists dis-
missed this prospect as highly improbable but the mere mention of the
possibility was enough to encourage Pedler and Davis to write, for publication by Souvenir Press in 1971, a work of fiction *Mutant 59: the plastic
eater*. This tale of disasters certainly encourages us to pursue the desirable
route of designing degradable polymers for use in throwaway items rather
than working to create plastic eating microbes which might escape from
refuse disposal sites to wreak havoc on our plastic civilisation.

Because the most conspicuous aspect of the plastic packaging waste was
litter, even though it only amounted to a fraction of the total weight of
plastic disposed, it is not surprising that the earliest technical solution
offered was to encourage the susceptibility of plastics to degradation by
sunlight. This was not a demanding step to take technically because most
synthetic polymers are intrinsically sensitive to photodegradation and a
variety of products were quickly on offer. The realisation that extra costs
were involved, set against the competitive nature of the packaging industry,
meant that these materials were not readily accepted especially when it
became fully appreciated that the photodegradation process took a significant time during which the object remained just as offensive to the eye.
Where it could be proved that actual harm was being done by the discarded

items the matter received more serious attention. The classic example is the well known six-pack ring binder for drink cans which had been confirmed as the accidental culprit in trapping wild birds. Certain states in the USA made legal history by being the first authorities to impose a legal obligation for the use of degradable plastics in the manufacture of six-pack binders.

As more thought was given to the scale and nature of the problem imposed on the community waste disposal activities by the inexorable growth of the plastics content of municipal solid waste (MSW) it became apparent that a variety of techniques needed to be available to cope with local conditions. Attempts to reduce the input of plastics by reducing the weight of polymer in individual packs has made a contribution, for example the use of thinner films made possible by the improvement in accurate gauge control, and the use of low density polyethylene/linear low density polyethylene (LDPE/LLDPE) blends or high density polyethylene (HDPE) films. It has also proved possible to intercept bulk waste packaging from goods inwards areas of supermarkets and redirect this material into recycling systems.

The direct recovery of polymers from MSW has proved to be technically extremely difficult and economically hopeless and this, in effect, leaves controlled landfill, incineration and composting as the only serious options.

Incineration has much to be said in its favour where the responsible authority has access to the substantial funds needed to provide and maintain equipment capable of dealing with the high calorific value waste, able to resist the corrosion caused by HCl from polyvinyl chloride (PVC) articles, and also scrubbing the exhaust gases to meet atmospheric purity standards.

Landfill remains, of course, the most widely used technique and has given rise to much argument about the prevailing conditions within the fill. As early as 1972 Wallhauser was reporting from landfill excavations in Germany that paper was surprisingly durable with newspapers readable after seven years burial. He also noted that many plastics items appeared to have undergone a degree of degradation. It is known that intense biological action, such as to maintain quite high temperatures, is to be found in many landfills and is associated with the anaerobic generation of such large amounts of methane that it can be tapped and used as fuel. It is equally clear that in the upper layers of the sites much of the younger material must be exposed to aerobic biological activity. It is also known that some landfills have been so effectively sealed and capped that, if the moisture content is low, then biological activity is minimal. It can be inferred that many landfill sites are able to biodegrade appropriate organic material and, accepting the serious nature of the growing problem of waste disposal, biodegradable plastics would be a beneficial introduction in MSW provided only that first they do not make the drainage water more toxic than it currently is, second they do not add to the problems of incineration, and third that they are compatible with schemes that may be developed for recycling plastic separated from MSW.

Composting in effect, is taking the aerobic layer of a landfill and dealing with it as a separate operation, often adding positive aeration and mechanical mixing. This technology is to be found in the early history of farming and gardening dealing with vegetable materials and some animal wastes. In more recent times it was mechanised and introduced for treating MSW usually by mechanically shredding the waste, after hand picking to recover items possessing market value such as textiles, then increasing its moisture content by adding sewage sludge before passing the mixture through a mechanical system which combined slow agitation with positive pressure aeration. Temperatures as high as 70°C were reached, most pathogenic bacteria were killed, and a substantial weight and volume loss occurred due to water evaporation and actual biological metabolisation of the organic content of the waste. Examination of the product revealed that the plastics items were still present apparently unchanged. Mechanical screening of the discharge after about five days transit through these plants yielded a modest amount of oversize material which would be sent to landfill, and a larger fraction which could, after maturing for a few months, be used to enrich ground for parklands or forestry sites. The capital cost and maintenance expense of these installations is rather high, nevertheless they have been installed in many places around the world. The USA has adhered to landfill operations plus, at one time, ocean dumping, but increasing pressure from concerned organisations has lead to many American municipal authorities introducing requirements for certain types of organic waste to be excluded from MSW and sent instead to fairly primitive composting sites where the material is simply stacked on hardstandings with free air access and, in some cases, regular turning of the heaps by mechanical equipment.

On consideration of the options it would appear that a biodegradable plastic that fitted the specification outlined above would make a positive contribution to the MSW problem subject to the further proviso that it did not contribute more than a 10–20% overall cost increase compared with the existing polyolefin materials.

In the early 1970s work was in progress at two non-commercial centres addressing this problem of creating a material that would be acceptable to the packaging industry on the grounds of cost, physical properties and toxicology. One such centre was the USDA laboratories in Peoria, Illinois, as part of a programme of work aimed at finding non-food markets for farm products. The Peoria biodegradable plastic project approached its target by adopting starch as the matrix natural polymer, gelatinising the starch by the action of heat and water, and then seeking synthetic polymeric additives which would make the mix processable on familiar plastics machinery. A certain technical success was achieved and US government patents were filed, but the process involved using a strong ammonia additive and the product film, with starch gel being the continuous phase, was much reduced in strength in the presence of moisture. In consequence of these technical matters, and perhaps because of the complexities associated with the licensing of 'public' patents, the work did not reach the marketplace for some years. The other centre of activity was at Brunel University in the UK where the complementary technique of using polyolefin polymers as the continuous phase with particulate starch additive as a filler was used. Extra additives ensured that the synthetic polymer phase oxidised on a controlled timetable and, as a result, the product was almost indistinguishable from the familiar polymer materials and could be processed on existing machinery without costly alterations. The industrial sponsor accepted the costs of patenting the original process and as a result the product reached the European market in 1974 in the form of LDPE shopping bags which are still in production today. The commercial history of both the Peoria and the Brunel work over the following years has been complex but these two types of product with their different properties and costs have both found markets and contributed in some modest measure to slowing the avalanche of disposable plastics items. It is also important to realise that the introduction of starch into polymer systems has started a diminution of the energy demands made by the packaging industry and also decreased the reduction of the world's non-renewable oil resource.

The emergence of a recognisable field of scientific activity around biodegradable plastics has encouraged the development of a wide range of new activities which have no connection with the original target of dealing with waste plastics. Studies of degradable materials for use in medical applications such as implants, and slow release systems for drugs have already borne fruit. Ideas have emerged for slow release pesticide and fertiliser systems in agriculture as well as mulch film systems which can cope with the need for parts of the film to be buried to retain the film in place. Technically successful methods have been demonstrated for controlled dosage in veterinary practice. We are also likely to see seedling planting systems made from biodegradable plastics as well as tree guards to protect saplings. The medical applications, in particular, are not as sensitive to the cost of the special polymers used so that it has become possible to consider synthesising materials such as polyesters and polylactic acids which would have been quite out of the question for disposable packaging applications. Polymers, such as polycaprolactone, which had been developed for other specialised applications were examined biologically because of their structure, and found to be biodegradable. Early microbiological research on bacterial metabolisms recording the presence of polymers granules in the cells of certain microorganisms was reinvestigated and the β-hydroxypropionate polymers were isolated on an industrial basis and shown to be biodegradable thermoplastics. These exotic materials all seem to fall in a price range of 4 to 10 times the price of plain polyethylene so that, although some have been evaluated for packaging applications, it is not surprising that this is for specialised packaging items such as in the cosmetics trade.

Recognising that much of the work on the various materials mentioned above has only been reported in diverse scientific publications, or in the shorthand of commercial literature, an attempt has been made in this book to bring together reviews and details of the commercially important materials on offer with as much information as can be gleaned on their processing technology and properties, including some detail on the business of recycling biodegradable materials. As in all competitive businesses, especially where there is current active technical development in progress, considerations relating to the protection of patent applications and also other commercial negotiations have not eased the task of the editor but he hopes, nevertheless, that the collection of specialists who have been persuaded to contribute will have put a good deal of the key information at the disposal of the many people who wish to enter into this field of work or who simply wish to understand its implications rather better than the efforts of well intentioned media writers may have made possible.

2 Chemistry and biochemistry of polymer biodegradation

A-C. ALBERTSSON and S. KARLSSON

2.1 Introduction

Biodegradation is an event which takes place through the action of enzymes and/or chemical decomposition associated with living organisms (bacteria, fungi, etc.) or their secretion products. Microbiological deterioration can be achieved by exo- and endo-enzymes or by products secreted biochemically or chemically from these. Macroorganisms can also eat and, sometimes, digest polymers and give a mechanical, chemical or enzymatic ageing. There are many different degradation modes that in nature combine synergistically to degrade polymers. Biodegradation might be better used as a term only when it is essential to distinguish clearly between the action of living organisms and other degradation modes (e.g. photolysis, oxidation, hydrolysis).

The accessibility of a polymer to degradative attack by living organisms has no direct relation to its origin and not all biopolymers are truly biodegradable. Complex macromolecules such as lignin are extremely inert while synthetic polymers with hydrolysable backbones, such as aliphatic polyesters, are accessible to the biodegradative action of esterases despite the usual specificity of these particular enzymes.

Chemical structure of the substrate is critically important for any enzymatic attack and the creation of a new kind of enzyme such as, for example, one which will degrade polyethylene, is probably not possible by the induction of biosynthesis. Enzymatic induction is just a triggering of an existing gene to produce a number of molecules of the corresponding enzyme. Thus a purely enzymatic degradation of a long straight olefin chain dependent on the enzymatic cleavage of the C–C bond should not be expected, since such an endoenzyme does not occur in nature. Oxidation is instead the critical initial step in the degradation of many rather inert organic molecules.

2.2 Enzymes

Enzymes are biological catalysts, with the same action as chemical catalysts, i.e. by lowering the activation energy they can induce increases in reaction rates in an environment otherwise unfavourable for chemical reactions, e.g. water at pH 7 and 30°C. Enzymes are among the most potent catalysts known, a rise in reaction rate of 10^8–10^{20} can often be observed in the presence of enzymes. All enzymes are proteins; a polypeptide chain with a complex three-dimensional structure. The enzyme activity is closely related to the conformational structure, even small changes in temperature, pH or osmolarity can result in changes in the conformational structure rendering the enzyme inactive. Their action is regulated, i.e. they can change from a period of low activity to periods of high activity, through the influence of hormones, pH changes or other factors.

The three-dimensional structure of enzymes with folds and pockets creates certain regions at the surface with characteristic primary structure (i.e. specific amino acid sequence) which form an active site. At the active site the interaction between enzyme and substrate takes place leading to the chemical reaction, eventually giving a particular product. Some enzymes contain regions with absolute specificity for a given substrate while others can recognise a series of substrates.

For optimal activity certain enzymes must associate with cofactors which can be of inorganic or organic origin. The inorganic molecules are metal ions, e.g. sodium, potassium, magnesium, calcium or zinc. The organic cofactors are also called coenzymes and they can vary in structure, some are derived from different B-vitamins (thiamine, biotin, etc.) while others are important compounds in the metabolic cycles such as nicotinamide adenine dinucleotide (NAD^+), nicotinamide adenine dinucleotide phosphate ($NADP^+$), flavin adenine dinucleotide (FAD^+), adenosine triphosphate (ATP), etc. Enzyme plus cofactor is called a holoenzyme while enzyme lacking cofactor is denoted an apoenzyme.

2.2.1 Enzyme nomenclature

All enzymes, except those few retaining historically important trivial names (trypsin, pepsin, etc.), are named according to rules adopted by the International Enzyme Commission. The names give the nature of the chemical reaction catalysed and also describe the substrate. All new enzymes end with the suffix -ase, but shorter names are often used as some enzyme names become very long, e.g. hexokinase for ATP: hexose-phosphotransferase. Typical enzyme classes, together with the reactions catalysed and the reactive bonds, are shown in Table 2.1.

Table 2.1 Reactions catalysed and reactive bonds of different classes of enzyme

Enzyme class	Reaction catalysed	Reactive bonds
1. Oxidoreductase	Redox reactions	$>C=O$
		$>C-NH_2$
2. Tranferase	Transfer of functional groups	One C-groups Acetyl groups
3. Hydrolase	Hydrolysis	Esters Peptides
4. Lyase	Addition to double bonds	$-HC=CH-$
		$>C=O$
5. Isomerase	Isomerisation	Racemases
6. Ligase	Formation of new bonds using ATP	$-C-O-$ $-C-S-$ $-C-N-$

2.2.2 Enzyme specificity

For enzymes with absolute specificity the 'key-and-lock' theory, which implies an unchangeable rigid conformation, is a plausable model. For enzymes with variable specificity the 'induced-fit' theory tries to explain its function. The initial contact between enzyme and substrate forms an optimal orientation at the active site giving good possibilities for maximal bonding (enzyme–substrate), often the cofactor induces these changes when binding to the enzyme.

2.2.3 Physical factors affecting the activity of enzymes

All enzymes are adjusted to a specific environment in which their activity and three-dimensional structure are optimal for the purpose. For human enzymes or enzymes isolated from human cells this environment is a water solution at pH 6–8, an ion strength of 0.15 molar (as is normal physiological saline at 0.9% NaCl) and a temperature of 35–40°C. An extremely small change in some parameter may render the enzyme totally inactive and sometimes even destroy it irreversibly. Other solvents than water, especially organic solvents, are also lethal to many enzymes but, on the other hand, there are enzymes active in quite extreme environments, e.g. in hot water springs or salt deserts.

2.2.4 Enzyme mechanisms

Different enzymes have different actions, some enzymes change the substrate through some free radical mechanism while others follow alternative chemical routes. Typical examples are biological oxidation and biological hydrolysis.

2.2.4.1 Biological oxidation Several enzymes can react directly with oxygen, the classical example being cytochromoxidase which is active in the respiratory chain. Oxygen has a special role in the metabolism of aerobic organisms.

In several cases oxygen is directly incorporated into the substrate. The enzymes can be hydroxylases (equation (2.1)) or oxygenases (equation (2.2)).

$$AH_2 + O_2 \longrightarrow AHOH + H_2O \qquad (2.1)$$
$$BH_2 \qquad B$$

$$AH_2 + O_2 \longrightarrow A(OH)_2 \qquad (2.2)$$

Hydroxylases are sometimes called monooxygenases and catalyse the insertion of a single atom of oxygen in the substrate A as a part of the hydroxyl group. The monooxygenases require a second reduced substrate BH_2 which simultaneously undergoes oxidation (i.e. dehydrogenation). Usually this second substrate is NADH (NADPH).

Oxygenases, also called dioxygenases, catalyse the insertion of the whole oxygen molecule into the substrate; sometimes the product is a dihydroxy derivative but more often the oxygen atoms are incorporated as part of a carbonyl (–CO) or a carboxyl (–COO–) grouping.

Yet another type of biological oxidation exists, namely the process where the oxygen molecule is not actually incorporated into the substrate, but rather it functions as a hydrogen acceptor (i.e. electron acceptor). Enzymes of this type are called oxidases, and one type produces H_2O (equation (2.3)) while another produces H_2O_2 (equation (2.4)).

$$AH_2 + \tfrac{1}{2}O_2 \longrightarrow A + H_2O \qquad (2.3)$$

$$AH_2 + O_2 \longrightarrow A + H_2O_2 \qquad (2.4)$$

One example of an oxygenase enzyme is that capable of catalysing the splitting of the aromatic structure producing two ($>C=O$) groups instead of the (–HC=CH–) group.

2.2.4.2 Biological hydrolysis. Several different hydrolysis reactions occur in biological organisms. Proteolytic enzymes (proteases) catalyse the hydrolysis of peptide bonds (equation 2.5) and also the related reaction hydrolysis of an ester bond (equation (2.6)).

$$
\left[\begin{array}{c} \underset{\substack{| \\ \text{H}}}{\overset{\text{H}}{\text{N}}} - \underset{\substack{| \\ \text{R}_1}}{\overset{\text{O}}{\text{C}}} - \overset{\|}{\text{C}} - \underset{\substack{| \\ \text{H}}}{\overset{\text{H}}{\text{N}}} - \underset{\substack{| \\ \text{R}_2}}{\overset{\text{O}}{\text{C}}} - \overset{\|}{\text{C}} \end{array} \right] + \text{H}_2\text{O} \longrightarrow
$$

(equation 2.5)

$$
\text{R}_1\text{-C-O-R}_2 + \text{H}_2\text{O} \longrightarrow \text{R}_1\text{-C}\overset{\text{O}}{\underset{\text{OH}}{\diagup}} + \text{R}_2\text{OH} \qquad (2.6)
$$

2.3 Chemical degradation initiates biodegradation

Scott [1] had already concluded in 1975 that the attack by microorganisms is often a secondary process. The step which determines the rate at which degradable polyethylene is returned to the biological cycle appears to be the rate of the molecular oxidation process, which reduces the molecular weight of the molecule to the value required for biodegradation to occur. Even in the absence of any biodegradative attack, the carboxylic acids produced may ultimately oxidise to carbon dioxide and water [1].

The initial step in oxidation should be compared with the initial step in degradation of lignin and other rather inert natural products such as rubber and gutta percha. It is also possible to compare this oxidation step with the degradation of polymer films implanted in the tissues of living animals [2] as well as the oxidation in a landfill.

Griffin [3] has attempted to summarise the evidence for the breakdown of synthetic polymers in the natural environment, especially under the influence of living organisms, by not only reviewing microbial and microbiotic impacts on both resistant and biosensitive polymers, but also presenting the penetration of hyphae of filamentous fungi in polyurethane and in starch/ethylene vinyl acetate (EVA) blends. Hyphal penetration in more or less solid structures is a common and generally known phenomenon both in soil colonisation and in saprophytic decay processes as well as in parasitic growth on soft or hard plant or animal tissues. Such hyphal penetration might deteriorate any possible inorganic or organic structure by simple mechanical effects in the microscale, by introduction of chemically active exudations in the surroundings (mostly organic acids), by oxidation–reduction processes and by the great variety of enzymes (many of them extracellular) which are the omnivorous scavengers of nature, leading to the total mineralisation of all organic and many inorganic litters. Attack of microbes, especially filamentous fungi, myxomycetes and bacteria like actinomycetes, on high molecular weight natural biopolymers of solid structure is a rule, not an exception, as is illustrated, for example, by mildew spreading on leather, textiles, fruit or animal skin.

In many landfills, especially at depths of 1 to 2 metres below the surface, or in deep soil burial especially in waterlogged land, anaerobic conditions can develop which then decrease the initial oxidation of polymers. Certain microoganisms can, however, utilise oxygen in chemically bound form from nitrate, sulphate, carbonate and fumarate anions and also, probably, from ferric ions in an anaerobic way without gaseous oxygen. In the absence of oxygen (at low redox potential) iron, steel and manganese can be polarised in the water or humid soil abiotically. Normally the hydrogen formed in this way adheres to the metallic surface as a thin layer and protects it from oxidative corrosion.

However, in circumstances when both sulphate anions and anaerobic sulphate reducing bacteria are present in the soil, as is often the case, the iron will be precipitated after cathodic depolarisation and sulphate reduction, with iron sulphide and ferrous hydroxide as final products.

In an earlier stage, sulphate will be reduced to hydrogen sulphide. Access to oxygen in a later stage of events can in turn lead to the conversion of ferrous hydroxide to the ferric form, which in turn can again be reduced by other bacteria (also reducing nitrates). With the passage of time and the diffusion of chemical products in the microenvironment, all these conversions might also occur in the vicinity of polyethylene structures and thus could provide the plastic surface with some oxygen in the transitory stage. Such oxygen, if generated, must in turn be able to contribute to an increase of peroxy- and carbonyl radicals in the inert polymer and give alkane utilising microorganisms an opportunity to attack the outer surface of the plastic structure.

Significantly oxidised polyethylene foils, when exposed to bacterial homogenates, can show the formation of terminal hydroxyl groups as well as intrinsic ketone and ester groups [4]. It has also been shown that cytochrome P-450 in mouse livers affects the formation of the oxidative groups on polyethylene[5].

Initial photoxidation has been demonstrated to have a profound effect on the degradability of polyethylene in soil even after a 10-year period of incubation [6]. Polyethylene without prior photoxidation degrades by some 0.2% during a 10-year period, while polyethylene photoxidised for 42 days before incubation in soil degrades by about 2%.

Biopolymers are the building stones of the intricate mechanisms of living cells, with a versatility copied by us only in a few cases (e.g. surgical polymers). On the other hand, biopolymers can also be recalcitrant, especially in the absence of oxygen, or because of molecular inertness, like some lignins and the higher alkanes (kerosenes, ozokerite).

Thermal rearrangements, especially reductive ones, will provide us with recalcitrant products from biopolymers such as asphalt, rubber, particular proteins (leather, casein) and cellulose derivatives eligible for a diversity of processing technologies. Thus the distinction should reflect the differences between molecular structures rather than the forces and geological time of production. The necessary enzymes for anabolism (biosynthesis) and catabolism (biodegradation, mineralisation) have probably already been provided by nature to ensure recycling of the basic elements of life (C, H, O, N, S, P, etc.) and only rather small modifications are considered by some scientists to be feasible through gene mutations, if one believes in the chances for spontaneous further development of the biosphere in contrast to human inventions. Alteration in a gene — a change in the heredity code — is the only way to create a new enzyme according to our present knowledge based on Beadle's 'one gene – one enzyme' theorem, but this 'new enzyme' must still follow the laws of chemistry with respect to atomic forces and molecular architecture.

2.3.1   The biodegradation of polyolefins

The mechanism for the biodegradation of polyethylene was presented in 1987 [7]. The mechanism shows similarities with the typical β-oxidation of fatty acids and paraffins in man and animal. An initial necessary abiotic step is the oxidation of the polymer chain; once hydroperoxides have been introduced a gradual increase in keto groups in the polymer is followed by a decrease in keto groups when short chain carboxylic acids are released as degradation products to the surroundings [8–10].

The enzymatic inertness of polyolefins, especially polyethylenes, can be interpreted by three different aspects of enzymatic activities:

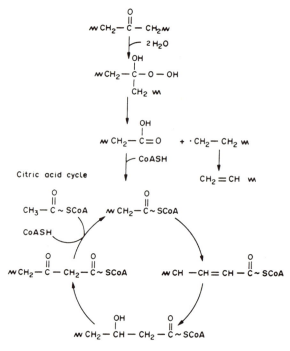

Figure 2.1 The biodegradation mechanism of polyethylene.

- Enzymatic degradation through β-oxidation of high alkanes, beyond $n\text{-}C_{44}$, is evidently hindered sterically. Thus C–C endohydrolysis in the longer straight or branching chains would be necessary in order to break up the polymers to lower oligomers before chain-end oxidation.

- No enzyme seems to exist in nature capable of splitting the C–C bond inside a normal paraffin chain (a chain at least less than the chain length of C_{44}), similar to the manner in which α-amylases attack cellulose.

- No 'endo-C–C-hydrolysis', cleaving the C–C branches, is known (this would have to operate with the molecular geometry involved in the functioning of amylo-1,6-glucosidase in cellulose).

Generally there are few carbon–carbon hydrolases capable of acting on carbon–carbon bonds and the enzymes described so far mostly catalyse the hydrolysis of 3-oxo-carboxylic acids. Consequently in the case cited of the degradation of medium chain alkanes, none of the additives can ever have stimulated the production of the degradative enzyme by so-called polymer influence and thus contributed to the metabolisation of low molecular weight products by other microorganisms (e.g. bacteria).

However, the combined effect of an abiotic oxidative step with consequent biotic action will be a slow but definite and progressive mineralisation.

The value of accurate observations of minute cumulative degradative processes caused by living organisms should be stressed here, the effect of which will in nature always be complementary and additive to ageing, weathering and other unavoidable abiotic impacts also of a cumulative type.

Steric hindrance is used here merely as a generalised expression of the influences at work preventing a substrate from reaching the catalytic groups in the active site of the enzyme and thus preventing it from forming an enzyme–substrate complex and being transformed, in this case oxidised and/or hydrolysed. With an increasing number of carbon atoms in the paraffin chain, both melting point and hydrophobic character will increase and make it more difficult for the substrate to fill the crevice of the active site in the large molecule of the enzyme, and thus establish contact with the polar groups, the anchoring contact in the crevice, which is necessary to establish an enzyme–substrate complex, a prerequisite for enzyme action.

2.3.2   Hydrophobicity influences biodegradability

Hydrophobicity is often regarded as a major obstacle to microbial (biodegradation) attack on polymers. Addition of surfactants in degradation studies with polyethylene showed a considerable increase in the biodegradation rate compared with samples without surfactant [11]. Ongoing studies incorporate surfactants in polyethylene showing promising results as observed by attenuated total reflection infrared spectroscopy, differential scanning calorimetry (DSC) and scanning electron microscopy (SEM) [12].

Even if the rather inert polyethylene is not attacked by enzymes in a first step there is an influence of living organisms and their secretion products.

It has been reported that *Bacillus cereus* (hay-bacterium) with a hydrophobic surface adheres most readily to hydrophobic surfaces such as silicone, polytetrafluoroethylene (PTFE), ethylene–propylene diene tercopolymers and low density polyethylene (LDPE) [13].

2.4   Hydrolysis of synthetic biodegradable polymers

Well known synthetic hydrolysable polymers are polyesters, polycarbonates, polyanhydrides and poly(amino acids). Many scientists are today looking for new possibilities using traditional natural polymers like polysaccharides, proteins and lipids. A special interest is focused on polyhydroxybutyrate and its copolymers. Also older materials like Pullulan, cellulose acetate and starch are being tested again as well as the synthetic polyvinylalcohol (PVOH).

Hydrolysable polymers, such as the polyesters, are often more prone to degradation but at the same time are often less suitable than hydrophobic polymers for many technical applications.

Aliphatic homopolyesters, e.g. poly(tetramethylene adipate) (PTMA) and block copolymers such as poly(ethylene succinate)-b-poly(ethylene glycol) (PES/PEG) and poly(ethylene succinate)-b-poly(tetramethylene glycol) (PES/PTMG) [14] have been synthesised and their subsequent hydrolytic degradation studied in a pseudocellular fluid buffered at pH 7.3 and maintained at 37°C. The materials obtained showed thermoplastic elastomer behaviour, the degradation rate depending on the polyether composition.

Poly (β-propiolactone) has been degraded in buffered salt solution (pH 7.2) at 37°C [15, 16]. Oriented fibres and non-oriented fibres show different degradation properties, especially with regard to the changes in mechanical properties. The changes in tensile strength are slower for the oriented material than for the non-oriented.

Polymerisation of 1,5-dioxepan–2-one gave a poly(ether ester) with amorphous properties, implying that it might be useful as an amorphous block in copolymers possessing elastic properties [17] and it should also be a candidate for hydrolytic degradation.

Aliphatic polyanhydrides degrade hydrolytically within a few days while aromatic polyanhydrides can degrade slowly over a period of several years. Recently, a new synthetic route for producing linear poly(adipic anhydride) by use of ketene gas has been presented [18]. This synthetic route has the advantage of avoiding formation of acetic acid, which could encourage the reverse reaction. Polyanhydrides are useful in biomedical applications due to their fibre-forming properties. An increase in the aliphatic chain length between the acid groups not only increases the molecular weight but also notably improves the hydrolytic stability [19, 20]. Aliphatic poly-carbonates are polymers derived from the polymerisation of six-membered ring molecules. The linear macromolecule degrades to the monomers at high temperature.

Poly(trimethylene carbonate) has been synthesised using cationic and anionic initiators, giving high molecular weight compounds with a rubbery character at room temperature [21]. Since the material obtained is an aliphatic polycarbonate, it could be useful as a biodegradable polymer for medical applications which should display hydrolytic degradation.

References

1. Scott, G. (1975) *Polym. Age*, **6**, 54.
2. Oppenheimer, B. S. *et al.* (1955) *Cancer Res.*, **15**, 335–340.
3. Griffin, G. J. L. (1980) *Pure & Appl. Chem.*, **52**, 399–407.
4. Wasserbauer, R., Beranová, M., Vancurová, D. and Dolezel, B. (1990) *Biomaterials*, **11**, 36–40.
5. Beranová, M., Wasserbauer, R., Vancurová, D., Stifter, M., Ocenáskova, J. and Mára, M. (1990) *Biomaterials*, **11**, 521–524.
6. Albertsson, A-C. and Karlsson, S. (1988) *J. Appl. Polym. Sci.*, **35**, 1289–1302.
7. Albertsson, A-C., Andersson, S. O. and Karlsson, S. (1987) *Polym. Degrad. Stab.*, **18**, 73–87.
8. Albertsson, A-C. and Karlsson, S. (1990) *Prog. Polym. Sci.*, **15**, 177–192.
9. Albertsson, A-C. and Karlsson, S. (1990) In *Agricultural and Synthetic Polymers* (eds G. Glass and G. Swift), ACS Symposium Series No. 433, p. 60.
10. Albertsson, A-C. and Karlsson, S. (1990) In *Degradable Materials* (eds S. A. Barenberg *et al*), CRC Press, Boca Raton, p. 263.
11. Karlsson, S., Ljungquist, O. and Albertsson, A-C. (1988) *Polym. Degrad. Stab.*, **21** 237–250.
12. Albertsson, A-C., Sares, C. and Karlsson, S. (1993) *Acta Polymerica*, **44**, 243.
13. Husmark, U. (1993) *Packmarknaden*, **3**, 34–35.
14. Albertsson, A-C. and Ljungquist, O. (1990) *Acta Polymerica*, **39**, 95.
15. Mathisen, T. and Albertsson, A-C. (1990) *J. Appl. Polym. Sci.*, **38**, 591.
16. Mathisen, T., Lewis, M. and Albertsson, A-C. (1991) *J. Appl. Polym. Sci.*, **42**, 2365.
17. Mathisen, T., Masus, K. and Albertsson, A-C. (1989) *Marcromolecules*, **22**, 3842.
18. Albertsson, A-C. and Lundmark, S. (1990) *J. Macromol. Sci. Chem.*, **A27**, 397.
19. Albertsson, A-C. and Lundmark, S. (1990) *Brit. Polym. J.*, **23** 205.
20. Lundmark, S., Sjöling, M. and Albertsson, A-C. (1991) *J. Macromol. Sci. Chem.*, **A28**, 15.
21. Albertsson, A-C. and Sjöling, M. (1992) *J. Macromol. Sci. Chem.*, **A29**, 43.

3 Particulate starch based products

G. J. L. GRIFFIN

3.1 Development of the technology

In 1972 this author initiated a project at Brunel University, at the request of the UK company Coloroll Ltd, seeking an economical method of conferring a degree of paper-like texture to low density polyethylene (LDPE) extrusion blown film. This was seen as an extension of a lengthy series of projects on special fillers in thermoplastics. The sponsoring company had an excellent reputation for its decorative paper shopping bags and, responding to the market demands for plastic bags, it decided to change to the use of plastics for manufacturing bags but remained convinced that the texture of paper was attractive to its customers. The economic constraints that applied to any medium sized firm in the plastics film business ruled out the possibility of developing an alternative polymer and those 'plastic papers' available at the time were far more expensive than paper itself or LDPE film. This meant that the only possible approach was to identify a particulate filler for the LDPE which would increase the stiffness of the polymer and develop an appropriate surface texture. Reviewing the fillers routinely available to the plastics industry suggested a variety of powdered minerals and a few fibrous cellulosic materials. The minerals were rejected on the grounds of weight, opacity and abrasive behaviour in the extruders. The fibres were rejected because of their disastrous effect on the rheological performance of the melt, it being quite impossible to use such porridge-like melts in the extrusion blowing process. This filler situation was not encouraging but, by chance, a new idea emerged from a parallel research programme active at Brunel in which natural whole grain starch suspensions were being used as a spherulite system emulating crystalline polymer films. Examination of the scanning electron microscope (SEM) pictures of the various starch particles gave rise to the idea of using starch as a particulate filler, encouraged by the knowledge that it was itself a crystalline organic polymer having physical and optical properties very similar to those of polyethylene.

Because compounding and film extrusion equipment was available in the University laboratories it was quickly possible to examine the feasibility of using starch as a particulate filler in polymer melts. Differential thermal

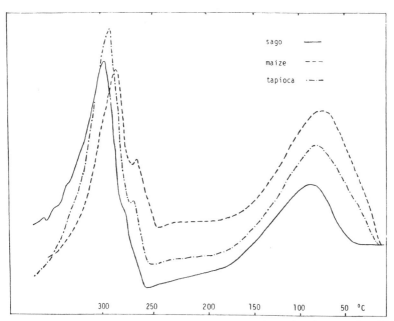

Figure 3.1 Differential thermal analysis plots from sago, maize and tapioca starches: 50 mg oven dried samples set up against alumina controls in Stanton DTA equipment.

analysis (DTA) measurements, as can be seen from Figure 3.1, showed that all the common starches displayed remarkable thermal stability up to an abrupt endothermic pyrolysis at around 265°C. Normal commercial starches contain between 12 and 16% water so that the early endothermic peaks visible on the DTA plots are entirely due to the evaporation of the water content. Trials with the extrusion of simple blends of commercial starches and LDPE granules dramatically confirmed the importance of drying the starch before mixing it with polymer melts. In the worst case the liberated steam returned through the granule pack in the feed section of the screw where it condensed and immediately gelatinised part of the incoming starch. With partly dried starch, say to 2% moisture content, the extrudate contained perfect starch granules and was free from dried starch gel, but the remaining water vaporised in the melt as it emerged from the die. The cooling and collapse of the steam bubbles created a silvery appearance due to the optical properties of the elongated lenticular voids left in the extrudate by the flattened bubbles. Reducing the water content to around 1.5% yielded a perfect visual appearance in the product but the adhesion between the starch and the polymer was minimal, as evidenced by the reduced physical properties of the extrudate. When the water content of the starch was reduced to < 1% the appearance and properties were optimal. Blown film could be produced reliably by making a masterbatch of the intensively dried starch and LDPE following conventional technology.

The only starches available commercially for this work in 1973 were potato, maize, wheat, cassava, sago and rice. The Coloroll company was only concerned with making high quality shopping bags with an average film thickness of 50 μm, soon to be reduced to 40 μm, so the large particle potato starch was not acceptable. The fine particle rice starch was too costly in Europe, and of the others, the most attractive option commercially was maize. With a particle size of about 14 μm maize was not the most attractive choice in terms of its effect on the tensile properties of the product films. The options for improving the strength of this starch/polymer composite were reviewed and experiments conducted to see if chemical bonding across the boundary could be achieved using reactive isocyanates. It soon became clear that the reactive route would require a substantial investment of time and money, give no guarantee of success, and would certainly add a cost penalty to the product. A decision was taken to follow the simpler route of improving the ease of wetting of the starch surface by the polymer melt. It was found possible to make the starch surfaces intensely hydrophobic by treating the powder with any of a wide variety of reactive silane products and, in this condition, the tensile properties of the starch/polymer blown films were improved by about 10%. At this point it was possible to consider the product for trials on commercial film blowing plants and, thanks to the cooperation of the UK company Viking Products Ltd, successful trials were carried out within the year.

By the early 1970s criticism of the plastics packing industry as a growing and uncontrolled source of pollution was gathering in intensity. It found its scientific focus in the reputed indestructability of the common packaging polymers when sheltered from sunlight. Having produced a novel composition, part natural and part synthetic, it was reasonable to examine the possibility that our starch/polymer composite might be more biosensitive than the synthetic polymer alone. It was clear from microscopy that the starch particles were in good condition and simply encapsulated in the polyethylene like flies in amber, suggesting that they would be unavailable to the usual starch digestion enzymes produced by many bacteria and fungi. Simple laboratory experiments in which film samples were suspended in α-amylase solutions and their starch content monitored by weight loss observations showed clear losses of part of the starch content leading to the scientifically uncomfortable conclusion that thin polyethylene films were permeable to quite large protein molecules. This observation was later pursued as a separate research project by Griffin and Nathan and the confirmation published in 1976 [1]. Encouraged by the early scientific indications, the Coloroll company filed a patent [2] and also encouraged the presentation of a technical paper in August 1973 [3]. Biological testing proceeded to trials in accelerated composting conditions modelled on the Dano system and reported in 1975 [4]. The interesting observation that degradation of the polyolefin component of the starch/ LDPE composites was observable in composting municipal solid waste (MSW) but not in simple garden soil burial lead to the realisation that some component of the MSW must be responsible. Chemical examination of LDPE films retrieved from composted waste at municipal facilities revealed that the active agent was unsaturated cooking oil which was selectively absorbed by the LDPE under the warm conditions of the Dano process. Up to 9% of rancid fat rich in peroxides could be found in these films and this could explain the degradative activity in terms of oxidative chain breaking followed by biological attack on the resulting fragments. The criteria for achieving biodegradation of polyethylene was set out very clearly by Hueck in 1974 [5] and it appeared that the presence of readily oxidisable unsaturated materials in the polymer formulation could achieve these criteria. Furthermore the possibilities of enhancing biodegradability by making mixtures of degradable and resistant substances were mentioned by Gorzynska in 1973 [6]. Accordingly the final formulation of the original degradable films marketed in 1974 by the Coloroll company emerged as a composition including about 7% of hydrophobic maize starch plus a small percentage of the pure oleic ester of octanol. Under moist conditions contact with earth of the discarded films would introduce sufficient transition metal contamination into the composite to trigger the autoxidation of the unsaturated fatty additive and, in turn, initiate the oxidative chain breaking mechanisms similar to those recognised as a consequence of photochemically generated peroxides.

Intense commercial interest in these products encouraged the Coloroll company to install its own twin screw compounding line making starch masterbatch for use on a new 45 mm Windmöller and Holscher film line using the graduated die system producing a bubble with thickened zones to reinforce the handle structure, these machines are shown in Figures 3.2 and 3.3. Over the following few years many millions of these shopping bags were produced, the plant was expanded to five production lines producing both normal and starch modified products and production continues to the present day. The company expanded into other decorative products, perhaps too rapidly for the general market condition to sustain its growth, and eventually it collapsed. Fortunately the packaging division, responsible for the plastic bag manufacture, found a new owner and remained intact with its original staff in place trading as Nelson Packaging Ltd. As a consequence of the company's collapse the 1973 patent, with several supporting patents [7–9] passed into the possession of the St. Lawrence Starch Co. of Toronto and, when that family business closed its starch plant, the activity remained with a subsidiary organisation now trading as Ecostar Ltd with its plant in Buffalo. The Coloroll company retained the right to manufacture the products independently for the

Figure 3.2 Original extrusion line installed at Coloroll Ltd. The M/B line is visible at the rear.

Figure 3.3 Original bag making line installed at Coloroll Ltd.

European market. The St. Lawrence Starch Co. declined to continue sponsoring the work at Brunel University and this author, the original inventor, continued the research as a private venture and, arising out of this work, a new patent was filed in 1987 [10] in which the fatty oil was replaced by a diene rubber, giving the same autoxidation sensitivity but also conferring improved mechanical properties on the product films and avoiding the odour problems and print adhesion problems associated with the fatty acid esters. It was also found desirable to include the transition metal catalysts as part of the polymer formulation as soon as the technology of managing the induction period was perfected because the acquisition of transition metal ions from earth or compost contact was a rather unpredictable phenomenon. The improved product was announced at the 1987 SPI meeting in Washington [11] and arising out of this disclosure came a licensing deal with the Archer Daniels Midland (ADM) company of Decatur, USA. This company made an important contribution to the manufacturing technology by bringing wet milling of maize, intense drying of starch, and the making of starch/polymer masterbatch all into the same plant. This vertical integration created a plant with attractive economic benefits many of which related to the energy conservation obtained as a result of introducing the hot starch directly into the mixing operation without the need for cooling and storing this intensely hygroscopic material. The activity was all the more impressive by being situated in the centre of the American corn growing region.

3.2 Current objectives

Twenty years of continuous research and development have been dedicated to the improvement and understanding of starch/polyolefin degradable plastics, from 1973 to 1987 at Brunel University and, from that time to the present date, the work has continued at the laboratories of EPRON Industries Ltd operating as the Ecological Materials Research Institute acquired from Brunel University. This activity has greatly advanced the techniques for converting the common packaging thermoplastics into degradable versions of themselves by the simple addition of a masterbatch additive. This process can now offer to the market materials which have controlled lifetimes whilst retaining most of the basic properties of the polymers used, and also doing this with minimal effect on product prices. All materials used in this technology are recognised internationally as being acceptable in terms of their low toxicity. Furthermore the formulations, sold in the USA and the UK under the trade name Polyclean™, can be converted by profile extrusion, blown film extrusion, film casting, bottle blowing, calendering, thermoforming or injection moulding processes using manufacturers' existing standard processing machinery.

Beyond the planned benefits of controlled degradability the Polyclean™ system offers the bonus of a satin texture which eliminates the blocking problem with thin films, and assists in maintaining accurate tracking of such films when traversing fast flexo printing machines.

Experience has shown that the masterbatch route is associated with a potential technical problem because it assumes that the converters are expected to monitor the quality of the final products from both the customary physical properties standpoint and also in terms of the degradability performance. It is not surprising that many such organisations will not normally possess the skills or equipment necessary for this aspect of quality control work. This situation has not been helped by the absence of any generally agreed standard testing procedures, although this should improve in the not too distant future (see chapter 6). An important objective of the development work over recent years has been the design of test procedures for assessing the life expectancy of the materials on the basis of short term testing methods because, obviously, lengthy biological procedures are quite inappropriate for product quality checking on a daily or even weekly basis. Now that it has been fully established that the breaking of the polyolefin molecular chains by autoxidation chemistry is the critical first stage in the biodegradation process it is evident that timing the onset of this chain breaking will yield a useful estimate of the overall degradation time. It is possible to monitor this chain breaking by using a variety of measurement techniques, as will be considered later.

Another important development objective has been the improvement in the physical properties of the products. This has entailed examining a variety of different polymer blends and also optimising the selection of appropriate diene rubbers for these formulations. Equally, if not more important, has been the pursuit of sources and types of starches better suited to the current preference for thinner plastic films. This generally means starches of smaller particle size and freedom from agglomerates, a topic which will be expanded later. In the competitive marketplace of commercial packaging materials the economics of degradable plastics must clearly have a dominant importance, given that the technical aspects are acceptable. Certain well appreciated principles of industrial economics apply to biodegradable plastics manufacture as well as they do to any other plastics operation. The principle of vertical integration is very relevant as has been seen in the ADM plant where in-house maize starch can be directed to the drier unit as required and immediately converted into masterbatch granules in a compounding line situated close by. It is interesting to realise that the original Coloroll plant started up in 1974 with a small twin screw compounding line sited next to the film extruder so that masterbatch could be made at convenient intervals as required. This would have been considered unorthodox because few film makers would have contemplated installing their own masterbatch compounding lines even though they are now often installing recycling lines using almost identical equipment. The more common situation would be an existing masterbatch specialist company buying dry starch from a starch manufacturer which is converted into masterbatch under license from the patent holders, and this masterbatch would then be sold on to film converters or other existing packaging manufacturers. It is clear that this method of organisation includes the largest number of profit centres, the maximum amount of material transport, and the poorest energy usage pattern. Yet another option, which has been considered but not extensively operated, would be for a polymer manufacturer to include starch and other additives in a special grade of degradable polymer which would then simply be an addition to their catalogue of grades and types. This latter scheme should be quite competitive and it would have the considerable attraction of ensuring that the quality control of the degradability aspect could be more readily maintained.

3.3 Relevant starch technology

The principal reasons for adopting starch as a filler in these biodegradable formulations have been set out earlier. Equally substantial arguments in favour of starch are the substitution by sustainable vegetable products of petroleum derived materials and the remarkably low energy consumption of the starch industry per tonne of product by comparison with other packaging materials. Data obtained from two major starch manufacturing organisations, incorporating government figures for the energy usage of the associated farming operations, provided the energy audit figures used to create the bar chart of Figure 3.4.

Further advantages in using this remarkable particulate polysaccharide are the surprising perfection of starch as a filler in terms of its particle shape and particle size distribution. To this can be added the benefit of using a filler with a high combined oxygen content possessing, therefore, a much lower calorific value than olefin polymers and giving less problem with the design of incineration equipment should the starch/polyolefin materials find themselves in the incineration route for MSW disposal.

The fact that starch is itself an organic polymer means that its density is of the same order of magnitude as the polyolefins, the figure usually quoted in the literature is ~ 1.5 g ml^{-1} but measurements by Linero [12] on intensely dried maize starch have produced density values as low as 1.25 g ml^{-1}. Costing calculations in the plastics film trade, being based on the idea of purchasing materials by weight and selling product by area, must obviously benefit from the 16.7% density reduction. The physical explanation, which is clear from Linero's work, resides in the case hardening of the starch particles as they dry which forces the volume contraction

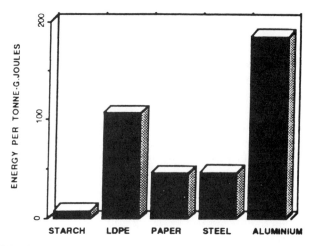

Figure 3.4 Total energy demand in the production of maize starch compared with other packaging materials.

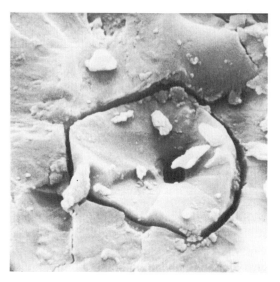

Figure 3.5 Fractured resin-embedded dry maize starch particle showing central void; magnification × 3500.

to take place from the inner hilum of the particles resulting in the formation of internal voids, as seen in the SEM picture of a fractured single particle in Figure 3.5, into which the viscous polymer melts cannot permeate.

It is also scarcely surprising, although not widely appreciated, that dry starch particles have a refractive index close to that of the common packaging polymers [13]. In those cases where the refractive index of the polymer matrix can be adjusted by manipulating plasticiser types and concentrations, or comonomer ratios in copolymers, film samples have been produced which are almost glass clear.

As was shown in Figure 3.1, the thermal stability of all the familiar commodity starches is virtually identical and therefore the basis for choosing a starch type for use as a filler for thermoplastic compositions is entirely a matter of particle size and cost. SEM pictures of the common commodity starches are given in Figures 3.6 to 3.8 and, for comparison, a potential microparticle starch from the weed cowcockle (*Saponaria vaccaria*) is included in Figure 3.9. Care is needed in interpreting 'average particle size' data given in specifications for starches because, although many starches have an almost monodisperse particle size distribution quite unlike the usual mineral filler powders, some do show a pronounced double peaked distribution, this being especially pronounced for wheat and potato starches. Because small concentrations of large particles render the material useless in the making of thin plastic films it is the largest particles present that determine the limit of acceptability. For the same reason, when checking starch samples for quality, their examination by microscopy should be performed on samples suspended in oils, and not water, because they frequently contain agglomerates of particles bonded together in the drying process which may not disperse in the plastic compounding operation. When suspended in water for microscopy the bonding, probably due to dextrin-like impurities, will soften and the agglomerates disperse easily giving the observer a false impression of their suitability for use in plastics compositions. Materials have been prepared using rice starch which, with an average particle size of 5 μm and a narrow size distribution, gives strong film products with the possibility of operating at film thicknesses close to 12 μm. Rice starch, however, is very costly on the European market and is also curiously prone to the agglomeration effect in its manufacturing process. Trials have been conducted with many different types of starch, some having particle sizes below 2 μm such as Taro root starch and cowcockle seed starch (*Saponaria vaccaria*).

3.4 Manufacture of masterbatch

A unique feature of starch, separated from maize seed by wet processing without harsh mechanical treatment likely to disrupt the natural particles, is that its particle size distribution is generally monodisperse and certainly does not contain any very fine fraction as is usually found in the mechanically ground powders commonly used as fillers for plastics. As a consequence it is easy to disperse dry starch in polymer melts and the process calls for good mixing rather than high shear dispersing. It was for this reason that the early large scale mixing trials were carried out using Buss KoKneader machines which proved to be excellent for the purpose. Care

Figure 3.6 SEM study of wheat starch; magnification × 525.

Figure 3.7 SEM study of rice starch; magnification × 2520.

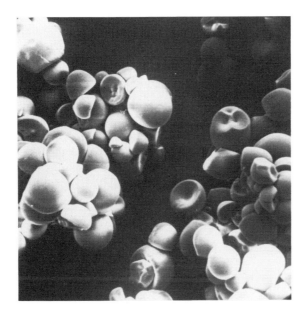

Figure 3.8 SEM study of tapioca starch; magnification × 500.

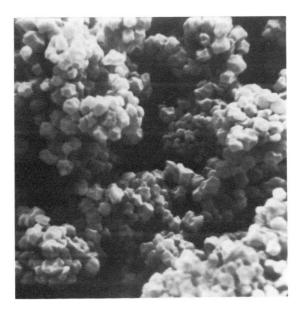

Figure 3.9 SEM study of *Saponaria vaccaria* starch; magnification × 2625.

must be taken to avoid moisture pick-up by the product and it was concern to protect the product that further encouraged the use of the Buss machines because face cutting facilities with air cooling and transport were available as standard items. It has subsequently been found that most twin screw extrusion compounding systems will handle these products and it is even possible to use conventional strand dies and cooling baths provided that the air knife arrangements for removing liquid water from the emerging strands are well designed and operating correctly. Excellent results have also been achieved using continuous internal mixer machines such as those made by the Farrell Bridge Company. When manufacturing masterbatches containing unsaturated fatty acid esters, or the more usual natural triglyceride oils such as rape oil, it is convenient to dose this minor ingredient into the compounding line by using a calibrated microdosing pump delivering into the root of the main feed hopper. This avoids the necessity of injecting the oil at high pressure directly into the melt through a boring in the extruder barrel. When operations were started using the new (1987) patent in which the metal derivative catalysts were included in the masterbatch formulation, the Polyclean™ materials were made as a composite masterbatch with the catalyst additives residing in a separate granulate so that the metal compounds and the unsaturated ingredients were not fused into a homogeneous system until the actual process of final film extrusion took place. In other words the storage life of the masterbatch was extended almost indefinitely whilst the storage life of the product could be predicted with reasonable accuracy starting from the date of product manufacture and subject only to variations in the storage temperature.

3.5   Conversion technology

3.5.1   Processing precautions—moisture

Very few problems arise in the conversion of starch filled polymers when using the conventional techniques of thermoplastics processing, and difficulties reported are usually traceable to careless handling of the masterbatch leading to excessive moisture pick-up. Careful attention to reclosing storage containers and not leaving masterbatch in exposed hoppers for extended time periods will usually ensure that problems of bubbling or streaking do not occur. In the event that mishaps of this kind do happen they can usually be circumvented by adding a few percent of calcium oxide containing masterbatch whilst processing the material giving trouble. It is, of course, possible to dry the starch masterbatch material but, even using desiccated air self-regenerating granule drying machines the process is rather slow and the trick of using supplementary desiccant masterbatches is preferable. These desiccant masterbatches, based on very fine particle size pure calcium oxide dispersed in polyethylene, are now obtainable from several specialist commercial masterbatch sources.

3.5.2 Processing precautions—temperature

As mentioned earlier, the pyrolysis temperature of all the common starches is close to 265°C, and because most of the packaging thermoplastics are processed at indicated melt temperatures between 170 and 230°C, starch pyrolysis is not usually a problem. Obviously certain thermoplastics, such as the polyamides and polyethylene terephthalate (PET), cannot be used with starch fillers. Certain extrusion operations, in particular the extrusion coating of paper, normally operate at polyethylene melt temperatures of >300°C in order to achieve a sufficient degree of adhesion between the paper and the polymer. This temperature is quite out of the question for starch materials which is unfortunate because of the major tonnage of these coated papers used in a variety of packaging applications. Successful trials have, however, been carried out on paper coating lines using ethylene vinyl acetate (EVA) copolymers which can offer the required level of adhesion but will extrude at much lower temperatures.

Care should be taken in assessing the suitability of particular extrusion processes for running starch/polyolefin materials because, on very fast operating production lines, the use of rather narrow die gaps can cause elevation of the melt back pressure which, in turn, can give rise to melt temperatures at the screw tip, i.e. the metering section, which are far above the indicated melt temperature. It is worth describing the nature of an overheating mishap because, although the melt will foam, become yellow or even dark brown, and vapours will be emitted smelling strongly of burnt bread, there is no need for alarm because by simply substituting plain polymer for the masterbatch blend feed the pyrolysed material can be rapidly flushed out of the extruder and die head. Technicians who have had bad experiences extruding unplasticised polyvinyl chloride (UPVC) may panic when confronted with brown foaming extrudate and simply switch of the drive before dashing to the nearest exit, the worst thing to do when extruding starch compounds.

3.5.3 Rheological considerations

When making thin films the usual starch concentration is ~ 6%, derived from a masterbatch addition level of ~ 15%, and this quantity has little effect of the melt flow index (MFI) of the melt, as can be seen from Figure 3.10 which records measurements made on two different polymers at various starch concentrations. It is obvious that the addition of a solid phase to a polymer melt will change its rheological character but examination of the plot makes clear that, because the recommended masterbatch addition of 15% corresponds to a starch addition level of 6%, the effect of MFI value is extremely small especially with polymers of low intrinsic MFI. Thick walled tubes and heavy (3 mm) sheets have been extruded from high density polyethylene (HDPE) and polypropylene (PP) containing starch contents up to 40% without thermal problems provided only that the screw speeds were watched carefully. Some processors have

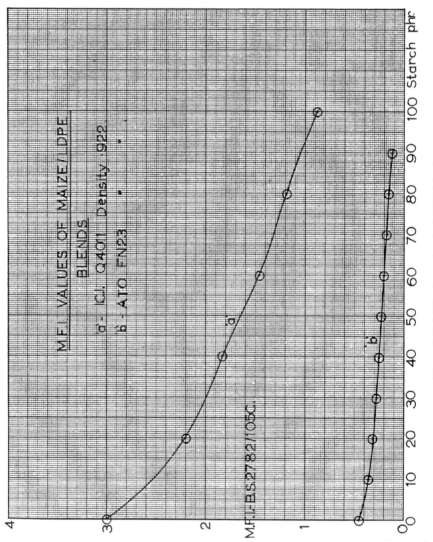

Figure 3.10 MFI values of maize/LDPE blends.

reported an increased output rate, as measured in weight per unit time, for a given film extrusion operation when changing over to extruding starch containing film. Although it is difficult to verify such reports and we know that there are density differences caused by the inclusion of the starch filler, the effect of spherical filler additions (glass ballotini) to polymer melts has been reported to have this rather surprising effect [14].

3.5.4 Cyclic conversion processes

Intermittent operations, such as injection moulding and, to a lesser extent bottle blowing, call for care because if the machine size is not well matched to the cavity volumes, then it is possible for the melt to be maintained at high temperature in the barrel and, possibly, in the hot runner system for excessive time periods which may lead to yellowing of the melt. Users of these starch products are well advised to adopt, as a standard practice, starting up the production using only straight polymer and then introducing the starch masterbatch once the conditions have stabilised. Conversely it is desirable to purge the lines at the conclusion of each shift and leave the machines to cool containing only straight polymer. Successful injection moulding trials have been run on modern fast cycling machines making disposable tableware with no thermal problems or difficulties in achieving fast cavity filling.

Many thousands of excellent HDPE bottles have been made using modern high speed extrusion blowing machines. Systematic measurement of the top loading deformation of such bottles, selecting a typical cylindrical detergent bottle shape to avoid confusing the data by the possible oblique distortion of a fancy bottle shape, have indicated quite convincingly that starch loadings of 12 to 15% increase the stiffness of the polymer far beyond the point that would be suggested by conventional mixing rules for composite systems. It is tempting to attribute this effect to the influence of the starch interface on the crystal morphology of the polymer but applying the conventional techniques for investigating morphology changes have failed to confirm this explanation. The effect, however, is obviously of economic interest because the critical top loading specification can be attained with a significantly lighter bottle.

3.6 Physical properties of products

The properties of any products made using masterbatch additive systems must be dominated by the properties of the base polymer, which would normally be provided by the manufacturer of the products. The standard Polyclean™ formulation recommendation is a masterbatch addition level of 15% for blown film products, although higher levels can be used in more solid products. This percentage is higher than the masterbatch addition levels used for pigment systems or other low level additives and will, therefore, make some modest differences to the physical properties of the products. In order to convey a useful idea of the influence of the Polyclean™ system on a typical commercial polymer, figures are quoted here for films made from one commercial film blowing grade of low density polyethylene (LDPE) (French ATO 1020FG24). This polymer has an MFI of 2 and density 0.918 g ml⁻¹ and has been used for degradation studies in the author's laboratories over the past ten years. It is typical of such materials, as offered by most polymer companies, but only with respect to the physical properties normally quoted on specification sheets. It should be remembered that the antioxidants used differ widely from one supplier to another and, although this will have negligible influence on the physical properties, it will certainly alter the degradation times and rates.

3.6.1 *Sample preparation*

All the films described here were made using Polyclean™ masterbatches checked immediately before use by ascertaining that the volatiles lost in a forced circulation oven after 20 minutes at 190°C amounted to less than 0.5% of the weight of the 10 g sample.

Typically a 500 g batch for test making is prepared by cold mechanical blending of 455 g of commercial polymer with 45 g of Polyclean™ master-batch. The extrusion line used was made by the Betol Co. (UK) and was fitted with a 25 mm diameter screw of 20:1 *L/D* ratio. The die used was 50 mm diameter and the product film was handled in a simple vertical tower with the customary cooling air-ring, nip rolls and constant torque wind up. Where possible the films were made to a bubble width of 250 mm and thicknesses from 30 to 150 μm as required with the screw operated at 60 rpm. In all cases the line was started up using base polymer only and the usual adjustments made to establish steady running conditions. At this stage the hopper was emptied and the prepared batch introduced, occasionally a colour tracer was introduced to delineate more sharply the boundaries between batches. The wind up onto card tubes was carried out at the lowest feasible tension in order to avoid leaving extension stresses in the cold film whilst stored on the rolls. As a further precaution the rolls were cut through along a radius so that the resulting stack of relaxed sheets, which were an ideal size for cutting samples, could be stored until tested.

3.6.2 *Physical testing methods*

The basic evaluation of product physical properties, as it relates to thin films, is generally restricted to making measurements of ultimate tensile strength and percentage elongation at break. This approach is adequate for general quality control of film products but in our work assessing the degradation properties of Polyclean™ based products we have found that a much better basis for comparison is to measure the area beneath the load/extension curve which, it will be realised, is a measure of the total energy consumed to break the sample. The elongation at break value is most sensitive to degradation of the materials, whilst the tensile strength is a far less sensitive indicator and it may even increase at the outset of the process due to initial crosslinking processes caused by the free radical chemistry in operation. Because of these considerations we have programmed the computer interface on our tensile testing machine to record integrated breaking energy as well as the standard measurements. Procedures otherwise followed are substantially those of BS2782 or equivalent ASTM procedures as appropriate. Test laboratory conditions were maintained at 23°C and 50% relative humidity.

To make a judgement on the effect of Polyclean™ addition on the physical properties of our 'typical' LDPE, data has been selected which compares the figures for the base polymer alone with film made containing 15% of the additive masterbatch. Because of the surface texture produced by the addition of starch based additives, it has been observed that substantial errors arise if the thickness of the films is measured in the conventional manner using a mechanical micrometer. This error can be as high as + 16% with thin films and, because this figure is used directly in calculating the tensile strength, the value obtained would be 16% below the true value. Accordingly in Table 3.1 a gravimetric thickness figure has been calculated using Method 512 of BS2782: part 5: 1970.

The nature of the factors influencing film thickness measurement and the basis of the BS2782:512 gravimetric thickness method and its limitations are clear from Figure 3.11, and results obtained from a study of the relationships between starch content and film gauge as measured by micrometer and gravimetrically are shown plotted in Figure 3.12.

3.6.3 Test results

The effects of Polyclean™ addition on the physical properties of polyethylene are not simple, some properties being improved and some worsened. With regard to the rather large scatter of results for testing thin plastic films to destruction, the physical effects of Polyclean™ addition will seldom influence product design considerations. There are, however, some effects of the starch filler which may be regarded as advantageous in particular applications, such as the dramatic reduction in blocking properties. It is also interesting to note that the gas permeability of thin polyethylene films

Table 3.1 Typical properties of film products

Test method	Number of samples	Control; base polymer	With 15% Polyclean™
Thickness (μm)			
Mean	5	98.3	120.5
Min.		96.0	106.0
Max.		101.0	126.0
Corrected mean		100.0	93.7
Grammage (g m^{-2})	5		
BS2782 method 512A			
Mean		92.2	88.3
Min.		88.4	86.8
Max.		94.0	90.4
Density (g cm^{-3})	3	0.921	0.942
BS2782 method 5090			
Burst strength	5		
(N cm^{-2}; 100 μm film)			
PFMS3/67			
Mean		12.0	11.6
Min.		12.0	11.2
Max.		13.0	11.7
Dart impact (g/F_{50})	20	404	360
BS2782 method 3520			
Tensile strength (mNm^{-2})			
BS2782 method 3268	5		
MD UTS Mean		10.3	8.7
" " Min.		9.2	7.8
" " Max.		11.2	9.4
TD " Mean		7.6	9.0
" " Min.		6.6	8.0
" " Max.		8.7	10.1
TD Yield St. Mean		4.9	4.9
" " Min.		4.8	4.8
" " Max.		5.0	5.0
Elongation at break (%)			
BS2782 method 3268	5		
MD UTS Mean		357	390
" Min.		280	334
" Max.		426	426
TD " Mean		469	591
" Min.		384	534
" Max.		536	663
Tear propagation (g/100 μm film)			
BS2782 method 360A	5		
MD Mean		122	145
Min.		118	124
Max.		174	162
TD Mean		239	272
Min.		224	235
Max.		272	329

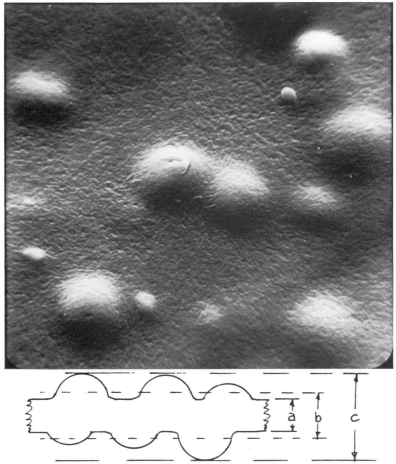

Figure 3.11 Top; maize/LDPE extrusion blown film surface. SEM study at magnification × 1100. Note leathery texture of PE stretched over starch grains. Bottom; three possible ways to express film thickness: (a) root thickness relates to load-bearing capacity but is difficult to measure, (b) gravimetric thickness, i.e. thickness of equivalent volume, (c) overall thickness as determined by micrometer.

is significantly reduced by the presence of starch particles, an effect which is recorded for dry air permeability in Figure 3.13. It must be appreciated that these effects will vary from one grade of polymer to another and also from one extrusion line to another. In critical applications, appropriate physical testing should be done before commencing manufacture and degradation properties should be checked, at least by thermal induction period measurement, in every case. Tabular data for all principal physical properties of our 'typical' polyethylene are shown in Table 3.1.

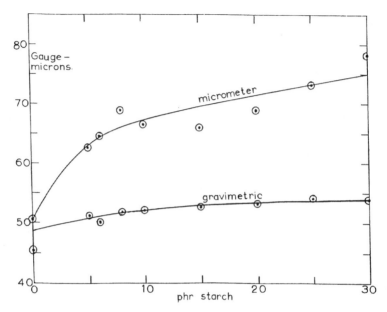

Figure 3.12 Relationships between starch content and film gauge measured by micrometer and gravimetrically.

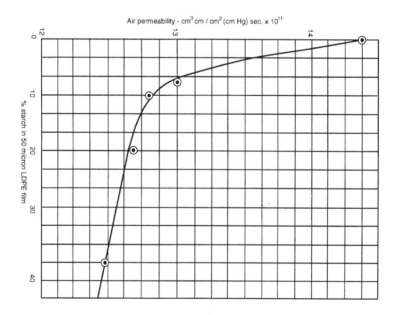

Figure 3.13 Effect of starch particles on dry air permeability of LDPE film.

3.7 Quality control testing of degradation

Great effort has been invested in selecting procedures for the accelerated testing of the degradation starch/polyolefin materials seeking, particularly, procedures which could form elements of a quality control procedure.

The complex processes by which the ethenoid thermoplastics, such as LDPE, HDPE, LLDPE and PP along with their numerous copolymers, degrade in the environment under biological influences are only partly understood but our appreciation of the mechanisms is improving and is discussed and reviewed in chapter 2. The biological effects of bacteria and fungi can only become significant after the molecular size of the polymers has been reduced by oxidation or photodegradation. Only the few percent of polymer waste that constitutes litter on the land or floating on natural waters is open to exposure from solar ultraviolet radiation, as also is the material used as agricultural mulch film. The majority of waste polymer from packaging applications goes directly into collection and disposal systems in which exposure to light is minimal. The starch/polymer/catalyst technology was developed to provide materials which have time controlled oxidation properties that would make possible the manufacture of products which retained their physical properties unchanged, regardless of contact with water or microorganisms and for which exposure to ultraviolet light was not essential to aid disposal. Biological action could be encountered in contact with the materials of the horticultural trade, the bacterial and fungal populations of many foods, or the microorganisms of composting or landfill operations.

It is, therefore, clear that the most important stage in the degradation process is the oxidative stage which follows the dormant or stable period and the Polyclean™ process controls this sequence of events by setting up a balance between the stabilising effect of the antioxidants present in commercial polymers and the autoxidation catalysts added as part of the Polyclean™ system. Each commercial polymer or copolymer will have different antioxidant behaviour and it is important that the balance between this effect and the autoxidant action is monitored to ensure that the system adopted for manufacturing maintains a life expectancy appropriate to the particular product and its working environment. It also has to be remembered that this life period should be such as to allow collection of parts of the waste plastic stream for restabilising the material in a recycling operation should local circumstances make this economically possible. As in the case of the physical properties reported earlier, we present here test procedures and degradation data as shown by 'typical' polymers that have been adopted as reference materials in the laboratories of our Research Institute.

Having established by the more academic and time consuming operations decribed earlier, that these materials degraded in the three clear stages of (1) the quiescent induction period, followed by the overlapping processes of (2) oxidative chain breaking linked to molecular weight reduction and (3) biological action as, encouraged by the stepping stone action of the degrading starch particles, microorganisms penetrate the brittle and crumbling plastic, it was clear that the future of the material could best be predicted by careful assessment of the time duration of the first stage, this being a reproducible technical operation not influenced by the chance variations of the disposal environments.

3.7.1 Autoxidation measurement

The simple procedure of suspending film samples in fixed temperature ovens with periodic sample removing for tensile testing enables a graph to be drawn up of physical strength against time. The maximum temperature which can be used for this work is usually about 70°C and the time requirement is typically between 5 and 14 days. Five sample strips are customary for each single tensile test and it will be readily appreciated that this method, although simple, is rather laborious and, of course, requires the availability of a sensitive tensile test machine. Some operators have felt satisfied with a simple assessment of oxidative breakdown by flexing the sample strips between the fingers and noting the point at which the strips cracked when flexed.

It is useful to discuss the measurement of the degradation process in terms of actual degradation life in such applications as the plastic bags used to collect garden waste in the USA because these bags are required to compost with the grass clippings and leaves and reach a specified state in a time fixed by the operation of the disposal site. The point in time after exposure at a given temperature at which a particular polymer film would be regarded as acceptable in a compost bag product is best determined by the measurement of physical properties because the primary requirement is that the bag should be disintegrating into innocuous fragments by the time that the ripe compost is ready for horticultural or agricultural use. As already discussed in section 3.6.2 the most relevant property parameter is the energy to break derived from the load/extension testing of aged samples and the test procedure that we have adopted involves ageing samples in incubators at three different temperatures (typically 50, 70 and 80°C, respectively). We have found it important to use incubators with forced air circulation and also to take care that the samples are held separate so that free circulation of air is maintained around them. Where samples are suspended in incubators, 80°C is regarded as the maximum temperature advisable to avoid deformation of the test pieces. This consideration does not apply to the gravimetric test procedures

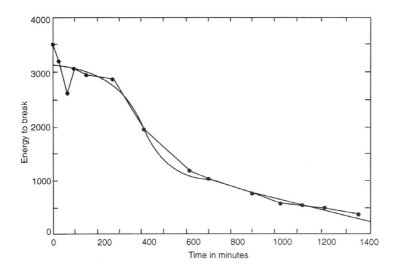

Figure 3.14 Strength loss (kJ) plotted against time at 70°C. Experimental points are plotted with a computer generated fitting curve overlaid.

where the very small samples remain supported in the platinum capsule of the instrument throughout the test. Higher temperatures are best avoided because the morphology of the solid polymer should remain unchanged.

As can be seen from Figure 3.14, which is a plot of breaking energy against time for one of our samples, the graph follows a complex form starting with a curious oscillation in which the breaking energy drops sharply before partly recovering during the first hour of exposure to heat. This phenomena is a type of annealing behaviour and has no connection with the oxidation process and should be ignored in interpreting the whole graph. The subsequent shape of the curve is a sigmoidal form in which the first downward inflection marks the end of the induction period and the beginning of the oxidation chemistry. Subsequently the strength drops rapidly until it enters a slow asymptotic stage towards zero strength. To make projections through tests at different temperatures it is desirable to have the lifetime of the polymer expressed as a reproducible single mechanical 'lifetime' figure and we have successfully fitted an equation to these sigmoidal curves which permits the derivation of a true 'half-life' time for each polymer formulation. These values can be plotted on a semi-log basis against temperature to give a straight line which simplifies the business of extrapolating half lives to other temperatures, a typical example is shown as Figure 3.15. Data for two different polymers are recorded and it is immediately evident that the slope of the degradation line is quite characteristic of individual polymer types. It is interesting to reflect that gathering

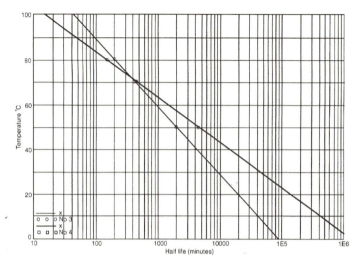

Figure 3.15 'Half life' time for two different polymer formulations plotted on a semi-log basis against temperature.

the data for preparation of Figure 3.15 involved making 140 individual tensile test operations.

It is also possible to use the same heat exposure regime but to monitor the changes by either repeated carbonyl index measurements using infrared spectroscopy, or by making repeated molecular weight measurements using gel permeation chromatography, an example of which is shown in Figure 3.16. These two techniques demand special skills and costly instrumentation. Work at our Research Institute over the past two years has shown that the autoxidation process can be monitored directly by following oxygen uptake which is manifested as a sudden increase in the weight of samples maintained at elevated temperature. What is being observed is the end of the induction period at which moment the chemical interaction between the antioxidant present in the commercial polymer and the autoxidation catalyst contained in the masterbatch has ended and the catalyst is free to start on the task of oxidising the polymer itself at which point more oxygen enters into combination with the polymer and the weight starts to increase. It is perfectly possible to make these measurements by suspending sample pieces in a constant temperature oven and removing them at regular intervals for accurate weighing. A typical plot of data obtained in this way is shown in Figure 3.17 where the now familiar sequence of events takes place, starting with a quick loss of weight associated with the loss of volatiles such as water from the sample, followed by the quiescent induction period and, finally, by the abrupt increase in weight as the induction period is finished.

By doing this same measurement using isothermal thermogravimetric

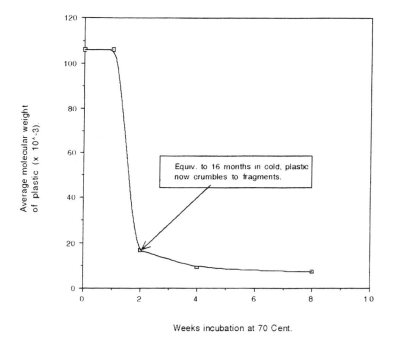

Weeks incubation at 70 Cent.

Figure 3.16 Molecular weight measurement using gel permeation chromatography.

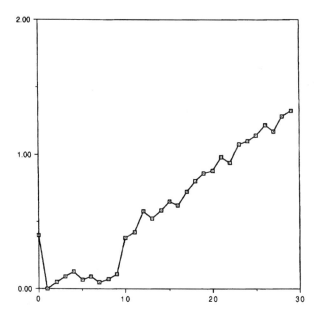

Figure 3.17 Plot of weight against time for a typical polymer, showing start and end of induction period.

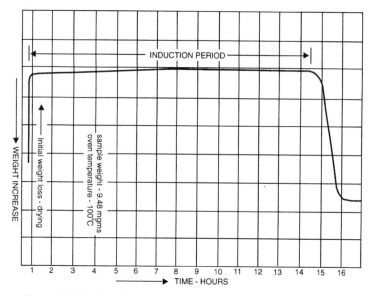

Figure 3.18 Isothermal thermogravimetric analysis of induction period.

analysis instruments the induction period measurements can be made on very small samples in a matter of a few hours with no operator action needed other than for cleaning the specimen pans and inserting the samples. The instrument output can be directed to a chart recorder or to a computer. A typical chart recorder record is shown as Figure 3.18.

Plotting the induction period semi-logarithmically against temperature yields a reasonable straight line suggesting that the action is following simple kinetics and this makes it possible to predict induction periods at lower temperatures with reasonable accuracy.

3.7.2 Biodegradation assessment

A great deal has been published on this subject and considerable confusion created by applying inappropriate test methods. The starch/polyolefin/ catalyst formulations are, in effect, not degradable by biological routes until they have entered the oxidative phase at the end of their induction period. It is true that a part of the starch content will be degraded when samples are exposed to burial in moist earth and this is helpful in establishing an adherent surface film of dextran gum which will assist in the biological processes which commence when material of reduced molecular weight becomes available from the oxidation mechanism. A typical record of molecular weight change after autoxidation is shown in Figure 3.16 and material below about 1000 Daltons is accessible to metabolisation by microorganisms. The truth of this action has been established by using

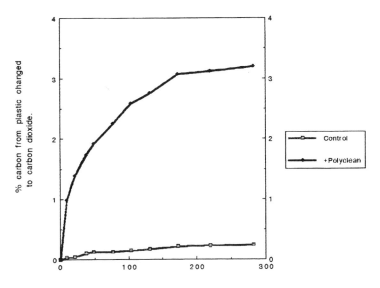

Figure 3.19 Monitoring microbiological degradation of autoxidised polyethylene using the radiocarbon labelling technique. Data from A-C. Albertsson.

specially synthesised polyethylene from ethylene gas containing radioactive carbon atoms. When films made from this polymer are first degraded oxidatively by prolonged storage at room temperature or, more quickly, by incubation and they are then exposed to cultures of microorganisms the carbon dioxide evolved by the metabolic activities of the organisms is found to be radioactive. This activity could only have come from the polymer by a biological mechanism. Figure 3.19 shows results obtained using this technique on a radioactive version of our 'typical' plastic (see section 3.6.1) and reproduced here by courtesy of Professor A-C. Albertsson of the Royal Institute of Technology, Stockholm. Details can be found in her publication [15].

3.7.3 Soil burial tests

Our own laboratory has adopted test methods more directly related to actual environmental conditions. In particular, soil burial tests with 50 mm wide strip samples buried on edge, to avoid waterlogging, in cultivated well drained but moist garden soil with the top of the samples ~ 50 mm below the surface. It was found necessary to protect such burial sites with wire mesh held on wooden frames above the soil surface to prevent disinternment of the samples by dogs, cats or rabbits. Our 'typical' film samples (see section 3.6.1) buried under these conditions in the south of the United Kingdom lose all mechanical strength after 18 months and are

Table 3.2   Alteration in tensile strength after period of burial

Time from burial	% Tensile strength
Initial	100.0
After 3 months	78.0
After 4 months	54.4
After 6 months	6.52
After 18 months	retrieval impossible

often impossible to retrieve for mechanical testing or visual inspection. Samples of this buried material were retrieved and tested for mechanical properties at intervals of 0, 3, 4 and 6 months when the percentage tensile strength changed as shown in Table 3.2.

The Ecological Materials Research Institute's laboratories in the UK are currently studying a variety of degradation test methods such as submersion of film samples clipped to stainless steel frames which can be submerged in the processing tanks of activated sludge sewage disposal works, accelerated soil burial tests in trays of garden soil maintained at appropriate humidity and temperature in a computer controlled environmental test chamber, and a version of the Sturm test procedures using infrared gas analysis apparatus to compare the CO_2 evolution from control systems of mineral nutrient solutions plus sewage inoculation with identical systems containing degradable plastic samples. A recent publication [16] by Johnson et al. reports the conclusion of an extensive trial of these granular starch/polyethylene films in actual municipal composting facilities and confirms the degradation process even in the second year of samples retrieved from the interior of the compost piles where it is known that the conditions are almost anaerobic.

It is widely recognised that the emergence of the new degradable plastics technology before the existence of established standard test methods has caused some problems in establishing good practices in the industry. This aspect of the subject receives detailed examination in chapter 6.

References

1. Griffin, G. J. L. and Nathan, P. S. (1979) *J. Appl. Polym. Sci.* **35**, 475–484.
2. Griffin, G. J. L. (1972) U.K. Patent 1,485,833.
3. Griffin, G. J. L. (1974) *Fillers and Reinforcement for Plastics* (ed. R. Deanin), ACS Symposium 134, pp. 159–170.
4. Griffin, G. J. L. (1976) Degradation of polyethylene in compost burial, *J. Polym. Sci. Sym.*, **57**, 281–286.
5. Hueck, H. J. (1974) *Internat. Biodet. Bull.*, **10**, (3), 87–90.
6. Gorzynska, J. (1973) *Przem. Ferment. Polny.*, **17**, (5) 36.
7. Griffin, G. J. L. (1978) US Patent 4,125,495.
8. Griffin, G. J. L. (1980) US Patent 4,218,350.
9. Griffin, G. J. L. (1982) US Patent 4,324,709.

10. Griffin, G. J. L. US Patent 4,983,651 (1991) and CIP (1993).
11. Griffin, G. J. L. (1987) *Degradable Plastic Films*, SPI Symposium on Degradable Plastics, Washington, DC. pp. 47–49.
12. Linero, G. O. (1979) *Ph.D. Thesis*, Brunel University, UK.
13. Griffin, G. J. L. (1985) Starch granules; properties and industrial applications, *New Approaches to Research on Cereal Carbohydrates* (eds R. D. Hill and L. Munck), Elsevier Applied Science, pp. 201–210.
14. Jones, H. L. (1975) *Influence of glass sphere/glass fiber reinforcement on flow parameters in the extrusion of H. D. polyethylene*, 30th. Anniversary Technical Conference, Reinforced Plastics Institute, SPI section 7c, pp. 1–4.
15. Albertsson, A-C, and Karlsson, S. (1989) *Prog. Polym. Sci.*, **14**, 74.
16. Johnson, K. J., Pometto, A. L., III and Nikolov, Z. L. (1993) Degradation of degradable starch-polyethylene plastics in a compost environment, *Appl. Environ. Microbiol.*, **59** (4) 1155–1161.

4 Biopolyesters

P. J. HOCKING and R. H. MARCHESSAULT

4.1 Introduction

Polyesters have been defined as polymers formed by the condensation of polyhydric alcohols, such as glycol or propylene glycol and polybasic acids, such as maleic or terephthalic [1]. Biopolyesters are polyesters derived from bacterial sources, and are exclusively based on hydroxyalkanoic acids. Usually these are β-hydroxyalkanoic acids, but some examples of γ-hydroxyalkanoic acids are known.

Although the ester function is widespread in natural products, acyl-linked polyesters have not received the widespread attention accorded to glycosidically linked biopolymers, namely polysaccharides. The discovery of biopolyesters leads to proposals that poly(hydroxyalkanoates) represent a new class of natural macromolecules, aligned with polyisoprenoides, polyglycosides, polynucleotides and polypeptides. The current interest in biopolyesters centres around their environmentally friendly nature, and opposition to extensive use of non-biodegradable plastics.

The simplest and most common member of the poly(β-hydroxy-alkanoate) family is poly(β-hydroxybutyrate) (Figure 4.1). This material is biosynthesized in a large number of bacteria as a storage material. Analogous to starch, the reserve material in plants, biopolyesters occur as sub-micron inclusions inside the cell. Present commercial development is based on fermentation technology, but genetic engineers are already cloning the genes which someday might enable the replacement of starch or lipids with biopolyesters in certain plants.

A number of publications have been written describing various aspects of biopolyesters [2–7]. The sections which follow describe the historical

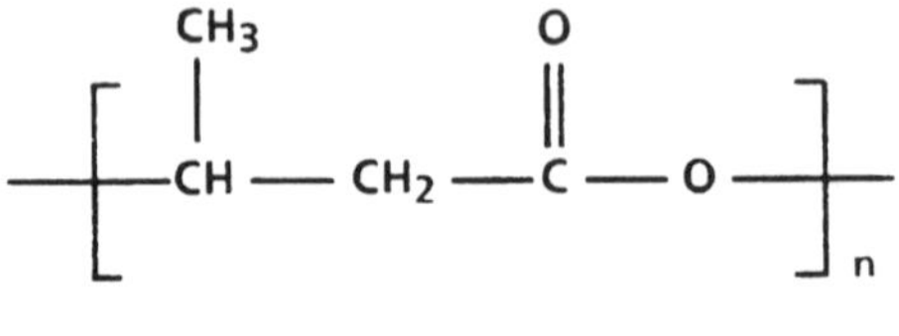

Figure 4.1 Chemical structure of poly(β-hydroxybutyrate).

development of knowledge of this polyester and the scientific and technical aspects of significance in its commercial development. At this stage of commercial development, biopolyesters are only available from fermentation technology, with current production approaching 1 000 000 lbs per year. The environmental thrust for recycling and clean technologies for waste disposal suggests that biopolyester production will increase significantly in the coming decades.

4.2 History

Poly-β-hydroxybutyrate (PHB) was mentioned in the microbiological literature as early as 1901 [8]. Detailed studies were reported by Maurice Lemoigne at the Lille branch of the Institut Pasteur, beginning in 1925 [9–12]. He observed granule-like inclusions in the cytoplasmic fluid of bacteria which were not ether soluble, as is normal for lipids. Using microscopic observations, saponification and acid numbers, autolysis, solubility and melting point variation with molecular size, and optical activity, he showed that this material was a polyester having the empirical formula $(C_4H_6O_2)_n$. He also accounted for the differences in melting point observed in two isolated fractions as due to differences in the degree of polymerization.

Over the next thirty years, PHB inclusion bodies were studied primarily as an academic curiosity. In 1952, Kepes and Péaud Lenoël observed that both polyester fractions isolated by Lemoigne were the hydrolysis products of a high molecular weight linear polyester with a melting temperature of 180°C, bearing a carboxylic acid group at one end and an alcohol at the other [13]. Weibull [14] made the correlation between the presence of intracellular lipid granules found in several bacterial strains and PHB. Early work on the polymer was reviewed by Williamson and Wilkinson [15], who were the first to report data on molecular weight and physical properties. Macrae and Wilkinson [16] observed that PHB accumulation could be increased by a limitation of nitrogen in the growth medium; following this, Merrick and Doudoroff [17, 18] examined the biosynthesis and enzymatic degradation processes of the polymer within bacterial cells [19]. Scientists came to the conclusion that bacteria store PHB as an energy reserve material, much as starch and glycogen are accumulated by other organisms [16, 20, 21].

In the late 1950s and early 1960s, Baptist and Werber at W. R. Grace and Co. in the USA began producing pound quantities of PHB for commercial evaluation. They obtained patents for the production and isolation processes [22–26] and developed articles such as sutures and prosthetic devices [27]. Their innovations extended to the use of unpurified high yield fermentation product in a plastic laminate [28]. However, their PHB fermentation yields were relatively low and their solvent extraction process was expensive. As well, the polymer produced was heavily contaminated with bacterial residues, making it difficult to melt process [29]. The project was abandoned and commercial interest lay dormant for over a decade. However, this work was truly pioneering both in terms of reporting on the plastic potential of PHB [24] and in proposing its use as a biocompatible material.

In 1968, ICI in the UK began developing single cell protein (SCP) technology, with the initial goal of developing a safe, highly nutritious food product for animal feed [30]. An SCP feed product called *Pruteen* was successfully developed, but the initial cost of the substrate for bacterial fermentation proved too high for large scale commercialization. The decision was made not to commercialize Pruteen, leaving the technology and facilities at hand to be applied to other projects.

By combining the expertise in large scale fermentation from the Agricultural Division with skills in polymer processing and evaluation from the Plastics Division [29], ICI was well prepared to tackle commercialization of PHB. The energy crisis of the 1970s was an incentive to seek naturally occurring substitutes for synthetic plastics. ICI found conditions such that the bacterium *Alcaligenes eutrophus* would produce up to 70% of its dry biomass as PHB. However, the mechanical properties of pure PHB showed no particular advantage over polypropylene, due to excessive brittleness. As oil prices stabilized, production costs for PHB remained higher than those for polypropylene, so the original idea of developing PHB as a general, high tonnage plastic was put on hold [30].

However, ICI had made a major advance in the production of PHB by patenting a procedure to produce copolymers of β-hydroxybutyrate (HB) and β-hydroxyvalerate (HV) [31] (Figure 4.2). This family of materials, known as PHBV or *Biopol*, has much improved properties over the original PHB, including reduced brittleness. Interest in developing bacterial polyesters was also increasing due to their biodegradability. ICI's work on PHB was made public in 1981 in a paper by Peter King [32], and in 1983 ICI and MTM (formerly Marlborough Technical Management) set up Marlborough Biopolymers Limited (MBL) to exploit Biopol [33]. In 1990, the first commercial product made of Biopol was launched in Germany as a biodegradable bottle for packacing a biodegradable shampoo by Wella [34].

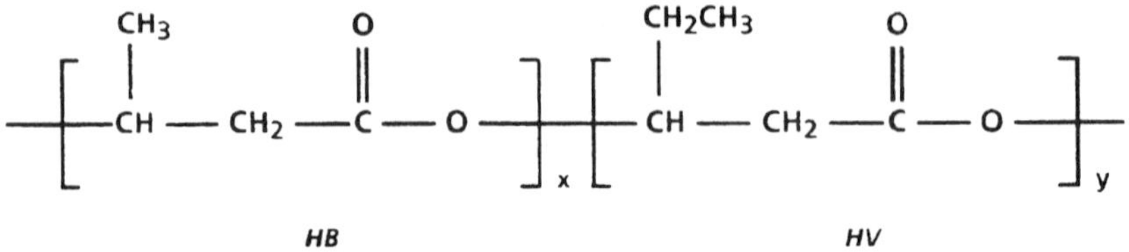

Figure 4.2 Chemical structure of poly(β-hydroxybutyrate-co-β-hydroxyvalerate).

4.3 Biosynthesis

Many aerobic and anaerobic bacterial species, under nutrient-limiting conditions with a sufficient supply of carbon, accumulate sub-micron inclusion bodies composed of poly-β-hydroxyalkanoates (PHAs). The most prevalent of these is PHB. PHB inclusion bodies are normally spherical, about 0.5 μm in diameter [18, 22, 23, 35], and play the role of an electron sink for the excess reducing power developed in an aerobic bacterium under conditions of oxygen limitation. In the case of nitrogen limitation, depolymerase enzymes are absent, accounting for the accumulation.

A scanning electron micrograph (SEM) of a section of bacterial cells packed with granules is shown in Figure 4.3. The water insoluble molecules of PHB exert a negligible intracellular osmotic pressure, making this polyester an ideal reserve material [19]. The amount of PHB in bacterial cells is normally 1–30% of their dry weight; however, under controlled fermentation conditions where nitrogen or oxygen are limited, some *Azotobacter* and *Alcaligenes* species can accumulate polymer up to 90% of their dry biomass [2, 22, 23, 36, 37].

Recent work has shown that PHB can be present in cells in a form other than as inclusion bodies; short chains (130–200 repeating units) of the polymer have been isolated from bacterial membranes [38–40] and also from a wide variety of plant and animal tissues [41, 42]. This PHB is found in small amounts (less than 0.001% of dry weight) complexed to lipids, proteins or salts of polymeric anions, suggesting a physiological function other than that of a reserve material [43]. Complexes of PHB with serum albumin or very low density or low density lipoproteins [42] may serve to modify the activity of the carrier protein or interfere with protein degradation. PHB in the plasma membranes of bacteria forms a complex with calcium and polyphosphate, postulated to be an exolipophilic–

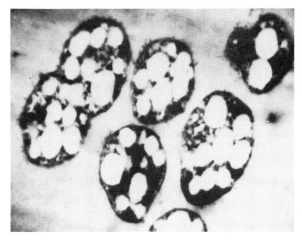

Figure 4.3 Scanning electron micrograph of sectioned *Alcaligenes eutrophus* cells packed with poly(β-hydroxybutyrate) inclusion bodies.

endopolarphilic helix about an inner framework of calcium polyphosphate which creates a channel through the membrane [40]. This complex is believed to play a role in regulation of intracellular calcium and phosphate [41], and in the uptake of single-stranded DNA [43].

Many studies have been made on the pathway of metabolic synthesis and degradation of PHB in microorganisms [17–19, 21, 44–55]. The pathway used by *A. eutrophus* (Figure 4.4) [47, 48, 52, 55] is the best known and probably the most widely distributed, and has been confirmed by experiments made using ^{13}C-labelled substrates and ^{13}C nuclear magnetic resonance (NMR) spectral analysis of the isolated PHB [54]. The building block is acetyl-CoA, produced from feedstocks such as glucose, fructose, sucrose, methanol, acetic acid or carbon dioxide–hydrogen mixtures [22, 23, 56]. A β-ketothiolase enzyme condenses two acetyl-CoA moieties to form acetoacetyl-CoA, which is reduced to *R*-β-hydroxybutyryl-CoA by an NADPH-dependent stereoselective acetoacetyl-CoA reductase. The third and last step of the biosynthesis is the action of the PHB synthase enzyme, which takes the *R*-β-hydroxybutyryl moiety and binds the β-hydroxyl group to the carboxyl end of a pre-existing PHB molecule to form an ester bond, increasing the chain length by one.

A wide range of molecular weights and melting points has been reported for PHB, depending on the method of isolation and the bacterial strain [57, 58]. Under mild isolation conditions such as solvent extraction or direct isolation of native granules [18, 22, 23, 59], high molecular weight polymers can be obtained, with weight average molecular weights ranging from a few hundred thousand to a million or more. An early study [35] combining granule size and molecular weight estimated that there were several thousand molecules per granule, a situation analogous to an emulsion polymerization. In a more recent study, the ratio of the molecular weight of PHB molecules to the total yield of PHB produced was used to predict that there were 18 000 polymerase molecules per cell, and that this number remained constant throughout the PHB accumulation period [60, 61]. A model involving chain initiation, propagation and transfer was proposed to account for the observed increase in the total number of chains during the PHB accumulation stage.

Some bacteria in natural environments such as estuarine sediments and sewage sludge can accumulate PHAs other that PHB [62–65]. By adding glucose and propionic acid to nitrogen- or phosphorous-limited cultures of *Alcaligenes eutrophus*, ICI has developed a process for producing one of these materials, PHBV, in large fermentors [31]. The ratio of HB to HV monomer can be varied by changing the glucose to propionic acid ratio [54]. By increasing the ratio of HV to HB repeating units, the melting points of the copolymers can be lowered [66], and their mechanical properties and thermoplastic characteristics are greatly improved [67].

NMR analysis of ^{13}C-labelled copolyester isolated from *Alcaligenes eutrophus H16* in culture media containing [1-^{13}C]propionate, [1-^{13}C]acetate and [2-^{13}C]acetate supports the mechanism of copolyester biosynthesis

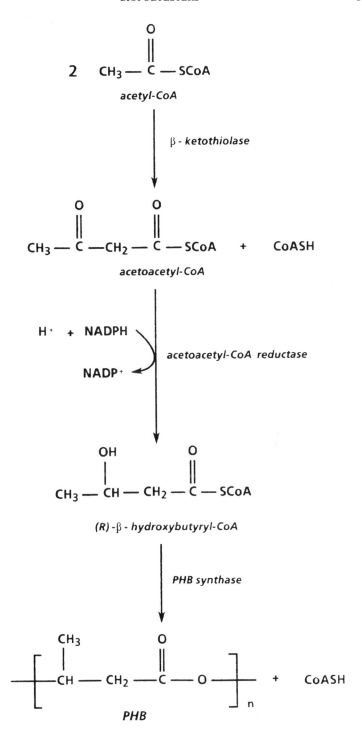

Figure 4.4 Biosynthetic pathway of poly(β-hydroxybutyrate).

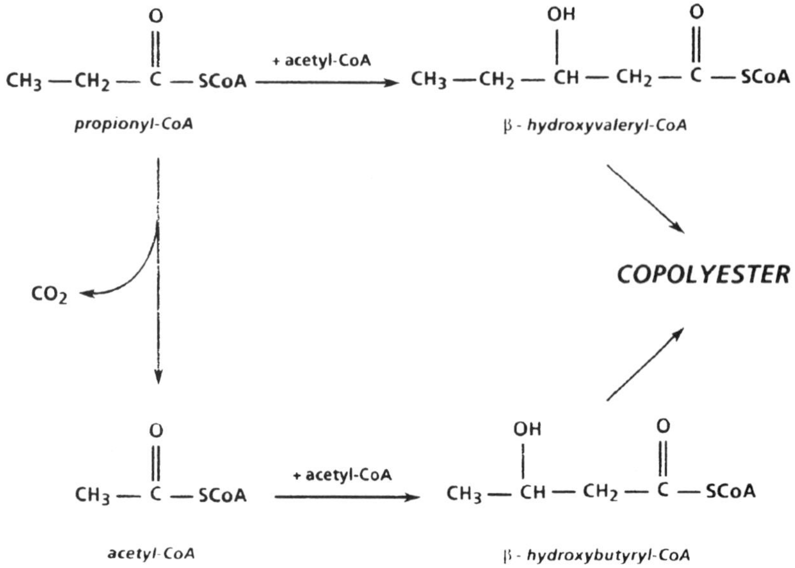

Figure 4.5 Biosynthetic pathway of poly(β-hydroxybutyrate-co-β-hydroxyvalerate).

shown in Figure 4.5 [54, 68]. When propionate is the sole carbon source in the culture medium, propionyl-CoA selectively eliminates the carbonyl carbon, forming acetyl-CoA. Two molecules of acetyl-CoA can condense to form β-hydroxybutyryl-CoA, or the acetyl-CoA can react with the propionyl-CoA to form β-hydroxyvaleryl-CoA. Finally, under the action of PHB polymerase, the β-hydroxybutyryl-CoA and β-hydroxyvaleryl-CoA copolymerize, forming a random copolyester. The ratio of HV to HB units in the copolyester increases with the concentration of propionate in the culture medium, suggesting that the ratio of β-hydroxyvaleryl-CoA to β-hydroxybutyryl-CoA increases with the concentration of propionyl-CoA in the cell. When both acetate and propionate are used as carbon sources, acetyl-CoA in the cell can be produced from either of them, decreasing the proportion of HV units in the copolyester.

Identification of the biosynthesis mechanism of copolyesters is important to allow control of their composition. When the culture medium contains glucose and propionic acid, the PHBV copolyesters isolated from *Alcaligenes eutrophus* have a range of compositions 0–47 mole% HV [29, 69, 70]. By using pentanoic acid as the sole carbon source for *Alcaligenes eutrophus*, Doi *et al.* [71] have obtained a copolyester with 90 mole% HV units; mixing glucose with the pentanoic acid in the culture media allows control of the fraction of HV units in the product.

Alcaligenes eutrophus is the bacterium presently used for commercial production of PHAs by ICI, but many other microorganisms accumulate these polyesters and can grow on carbon sources other than those for *A. eutrophus*. Some examples are *Pseudomonas oleovorans* [72–74], *Bacillus megaterium* [64], *Methylobacterium* [75], one species of *Aphanothece* [63], a cyanobacteria *Nocardia* [76] and *Pseudomonas cepacia* [77]. *Chromobacterium violaceum*, fed with a valerate carbon source under nitrogen limitation, yielded the first bacterial sample of 100% PHV [78], while a polymer containing 99% PHV was isolated from a *Rhodococcus* bacterium fed on valeric acid [79].

In 1989, Ramsay *et al.* [77] evaluated the possibility of using *Pseudomonas cepacia*, one of the most nutritionally versatile microorganisms [80], to produce PHAs by fermentation of inexpensive waste material. This bacterium is also resistant to the toxic effects of small carboxylic acids, which can accumulate locally in toxic concentrations due to imperfect mixing in the large fermentors used for commercial PHA production, killing more sensitive microorganisms [67]. *P. cepacia* grown in a mineral salts medium containing 0.5% fructose as the sole carbon source accumulated high molecular weight ($M_w = 5.4$ x 10^5 g mol^{-1}) PHB to 50% of the cellular dry weight [77]. Addition of propionic acid to the cultures yielded PHBV up to 30% of dry biomass weight; increasing the concentration of propionic acid increased the HV/HB ratio but decreased the total amount of polymer produced. These results suggest that *P. cepacia* could be a suitable alternative to *A. eutrophus* for commercial PHA production; although this microorganism can accumulate PHB on mixed waste residues, wastes that contain single carbon sources or lead to the production of only acetyl-CoA are required to obtain copolymer products of controlled composition.

4.4 Isolation

Once PHAs have been accumulated in bacterial cells, the cell walls must be broken apart and the polymer separated from cell debris. Care must be taken to avoid depolymerization during this process; early measurements of the physical properties of PHB isolated from various bacterial strains showed wide variations in molecular weights [57, 81], which later proved to be primarily due to the harsh methods of isolation used. Some degree of *in vivo* enzymatic hydrolysis may have also taken place before isolation.

Currently there are three approaches to isolating PHAs: solvent extraction, chemical digestion by sodium hypochlorite and selective enzymolysis. Efficiency is improved by centrifuging the fermentation broth, where the cell concentration is 50 g l^{-1}, to a concentrated cell paste which is then used as the starting point for polymer isolation.

4.4.1 *Solvent extraction*

Also known as the physical approach to PHA extraction, the solvent extraction method usually yields polymer with high molecular weight [57]. The PHA is extracted from the bacterial cell paste by dissolving in an organic solvent such as chloroform [7, 9, 11, 12, 82], methylene chloride [22, 23], 1,2-dichloroethane [83–85], 1,1,2-trichloroethane [84] or propylene carbonate [86]. The solution is filtered to eliminate bacterial cell debris, then the PHA is precipitated by slowly cooling the solution [86] or by adding a non-solvent such as methanol, diethyl ether or hexane [82, 84]. The polyesters can then be purified by redissolving in chloroform and reprecipitating with hexane or diethyl ether. The purity of the PHA recovered is improved if the biomass is pretreated with methanol or acetone prior to the solvent extraction, which increases the permeability of the cell membrane and can remove lipids and denature some low molecular weight proteins [82] that are usually extracted along with the polymer.

PHB recovered by solvent extraction is a white, highly crystalline powder of high molecular weight, while other PHAs with slower crystallization rates form films or aggregates which generally crystallize further with time. However, the method requires large volumes of solvents and non-solvents to extract and to precipitate the polymer. Accordingly, this method is not suitable for large commercial operations, and only recommended when high purity polymer is required.

4.4.2   Sodium hypochlorite digestion

Also known as the chemical approach, this method was first used by Williamson and Wilkinson [15] to isolate PHB granules from *Bacillus cereus*. The bacterial cells are treated for 30–60 minutes with a sodium hypochlorite solution, degrading and dissolving the cell wall and other non-PHA components but leaving the polyester granules seemingly intact. The polymer can then be purified by washing with diethyl ether or methanol to remove lipids. However, the high alkalinity of the system can cause chain cleavage, affecting the surface properties [15] and molecular weight of the polymer chains [57].

Several approaches have been taken to minimize the polymer damage caused by this method of isolation. Nuti *et al.* [87] showed that lyophilizing (freeze drying) *Azotobacter chroococcum* cells prior to hypochlorite treatment lowered the depolymerase activity, thus diminishing the decrease in degree of polymerization. They also observed that treating the cells with phenylacetic acid prior to lyophilization further stabilized the molecular weight. By carefully controlling the pH and digestion time, Berger *et al.* [88] were able to isolate high purity (95%) PHB of high molecular weight (600 000). Pretreatment with a surfactant increased both the purity and the molecular weight of the isolated granules [89]. Despite these advances, problems with this method remain: it is difficult to completely eliminate the traces of sodium hypochlorite from the polymer, and the sodium hypochlorite can pollute the environment.

4.4.3   Enzymatic digestion

This biochemical approach to PHA isolation was developed in 1964, when Merrick and Doudoroff [90] used lysozymes and deoxyribonuclease to solubilize the peptidoglycans and nucleic acids, respectively, of cells of *B. cereus*. The weakened cell walls were then ruptured by ultrasonic treatment to liberate the PHB granules in the buffered suspension medium. Granules isolated by this method are still capable of enzymatic synthesis and degradation of the PHB chains, and so are very close to their *in vivo* state. It was shown that 98% of the granule constituents were PHB and the remaining 2% were proteins with traces of lipids [91]. A modification of this method was recently used on *A. eutrophus* [92].

The most advanced version of the selective enzymolysis process is that described by Holmes and co-workers [93, 94], and shown in Figure 4.6. The process aims at the highest level of biomass production with the highest possible weight percent of PHA per dry weight of biomass (% PHA). The enzymatic digestion of cell components usually causes a sudden release of the nucleic acids (DNA) in the suspension medium, rendering the suspension highly viscous and virtually impossible to treat any further. To prevent this situation, a heating stage (steps 4 and 5) is used prior to the enzymatic digestion step to denature and solubilize the DNA. Following the pretreatment, several enzymes (alcalase, phospholipase, lecitase and lysosomes) are used, either in separate, sequential steps or, provided that the enzymes used will not interdigest each other, mixed together. The pH and temperature are carefully controlled so that optimum activity is obtained for the different enzymes.

The products obtained from this isolation procedure usually contain at least 90% PHA material, 1–3% peptidoglycans and 6–7% proteinaceous material. The PHA isolated is either submitted to an extra purification step by solvent extraction (as in the above description) or is spray dried, resulting in aggregates of granules 0.2 – 0.5 µm in diameter. The spray dried granules are partly crystalline and appear in the scanning electron microscope as grape-like granule clusters comparable in size to the spray droplets (Figure 4.7). Crystallization is initiated, at least at the granule surface, by the heat treatment (steps 4 and 5 in Figure 4.6). This stage of the process can be controlled to produce a desired level of depolymerization. Similarly, upstream steps in this purification process allow deliberate molecular weight breakdown, if desirable for certain product applications.

4.5   Properties

PHB homopolymer is a thermoplastic material, meaning that it is a resin that becomes highly viscous and mouldable at temperatures close to or

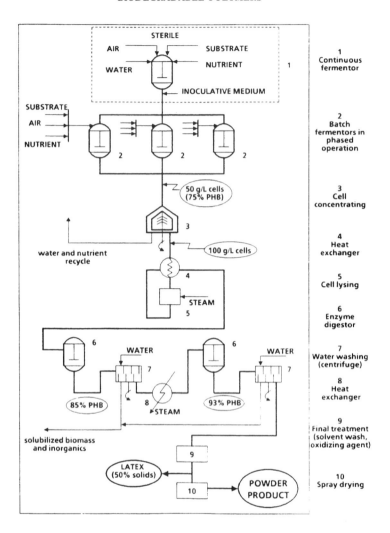

Figure 4.6 Schematic diagram for the isolation and purification of poly(β-hydroxyalkanoates) using selective enzymolysis. Redrawn and adapted from Marchessault *et al.* [117]. Reprinted by permission of Kluwer Academic Publishers.

above the melting point. Its properties are often compared to those of polypropylene (Table 4.1), as both polymers have similar melting points, degrees of crystallinity and glass transition temperatures (T_g). However, PHB is both stiffer and more brittle than polypropylene. The brittleness of PHB is largely due to the presence of large crystals in the form of spherulites, which form upon cooling from the melt; a hot-rolling treatment to remove cracks from within the spherulites can reduce the brittle character, allowing production of ductile films [95]. The materials also differ in

(a)

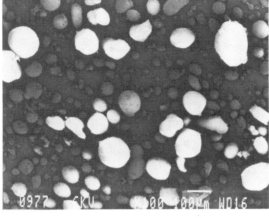

(b)

Figure 4.7 Scanning electron micrograph of poly(β-hydroxybutyrate-co-β-hydroxyvalerate) granule clusters created by spray drying: (a) low magnification; (b) high magnification, revealing the fine structure. Sample from Marlborough Biopolymers.

Table 4.1 Properties of PHB compared to those of polypropylene (PP)[30]

	PHB	PP
Crystalline melting point (°C)	175	176
Crystallinity (%)	80	70
Molecular weight (Daltons)	5×10^5	2×10^5
Glass transition temperature (°C)	− 4	− 10
Density (g cm^{-3})	1.250	0.905
Flexural modulus (GPa)	4.0	1.7
Tensile strength (MPa)	40	38
Extension to break (%)	6	400
Ultraviolet resistance	good	poor
Solvent resistance	poor	good

chemical properties, with PHB having a lower solvent resistance but much better natural resistance to ultraviolet weathering then polypropylene.

Early evaluations of PHB included the production of a number of mouldings, extrudates, films and fibres [29]. The properties of these were generally satisfactory but not spectacular. Packaging film had excellent gas barrier properties, being five times less permeable to carbon dioxide than was poly(ethylene terephthalate) (PET) [96], and was as strong as polypropylene film but not as tough as PET. PHB can be strengthened by addition of a glass fibre filling [32]; glass-reinforced mouldings were stiffer and tougher than similar nylon parts, but their heat resistance was not quite up to engineering specifications [96].

Properties are greatly improved in the PHBV copolymers. With increasing concentrations of HV units from 0–25%, there is a decrease in melting point (Table 4.2), increasing the size of the processing window within which the polymer can be melted without being degraded. The glass transition temperature also decreases, allowing use of these materials at lower temperatures without them becoming brittle and glassy. The steady decrease in Young's modulus (also known as the flexural modulus) indicates improved flexibility. The notched Izod impact strength also increases

Table 4.2 Physical properties of P(HB-co-HV) copolymers[3]

P(HB-co-HV) (mole % HV)	Melting point (°C)	Glass transition (°C)	Young's modulus (GPa)	Tensile strength (MPa)	Notched Izod impact strength (J m^{-1})*
0	179	10	3.5	40	50
3	170	8	2.9	38	60
9	162	6	1.9	37	95
14	150	4	1.5	35	120
20	145	−1	1.2	32	200
25	137	−6	0.7	30	400

*With 1 mm radius notch

with increasing HV concentration, indicating that the toughness of the material is increasing.

The biosynthetic origin of PHBV copolymers adds a number of interesting properties to these materials, relative to purely synthetic polymers. First, having been made by bacteria, these materials can also be degraded by bacteria, putting them amongst the few fully biodegradable, thermoplastic materials. Another result of its biosynthetic origins is optical activity, meaning that films or solutions of PHBV will rotate the plane of polarized light passing through them. This arises because the β-carbon in every monomer in a PHB or PHBV chain has an R absolute configuration; thus, the polymer is perfectly isotactic and can achieve a high level of crystallinity. These crystals have no centre of symmetry, resulting in a change in direction of average dipole moment if the crystals are deformed in a particular way [29]. This produces a polarization such that a surface charge is generated in response to deformation in shear modes; in short, the material generates a voltage when squeezed, or deforms when a voltage is applied to it [96]. This property is known as piezoelectricity, and is typical of many biological systems, but not usually of plastics [96].

4.5.1 Crystal structure

A number of research programs have focused on the crystal structure of PHAs. Wide angle X-ray diffraction (WAXD) studies [81] performed on oriented fibres and on single crystal stacks of PHB yielded a fibre pattern with a repeat distance of 5.96 Å, which required the chains to be in a helical conformation. The 50 Å thick lath-shaped lamellar crystals observed by transmission electron microscopy (TEM) are folded chain crystals having their fibre axis normal to the lamellar surface (Figure 4.8). This type of folded chain arrangement forms readily when dissolved PHB chains are precipitated from dilute solution, and is a common feature of the crystalline morphology of most synthetic polymers [97]. PHB has such a propensity for assuming this lamellar morphology upon solution precipitation that it was used as an identifying feature in certain studies [57, 98]. PHV forms square crystals corresponding to a different unit cell [99].

Later studies, including X-ray fibre diffraction [100], energy minimization analysis [101], and diagonal least-squares analysis [99], concluded that the crystalline PHB molecules are packed in an orthorhombic unit cell in a helical form, with a two-fold screw axis along the chain propagation direction. Although the first study [100] proposed right-handed helices, the latter two [99, 101] concluded that the helices were left-handed. The first reported [100] crystal lattice parameters are still valid, with $a = 5.76$ Å, $b = 13.20$ Å and $c = 5.96$ Å. Two chains of opposite polarity pass through the unit cell, in accordance with the $P2_12_12_1$ space group (Figure 4.9). Oligomers [102] of PHB adopt the same helical confor-

(a)

(b)

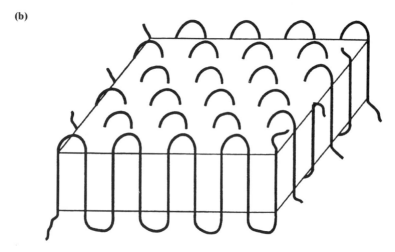

Figure 4.8 Folded chain lamellar crystals of poly(β-hydroxybutyrate). (a) Transmission electron micrograph of poly(β-hydroxybutyrate) single crystals precipitated from dilute triacetin solution. Inset is electron diffractogram from circled region. (b) Schematic of folded chain lamellar crystal. From Lauzier *et al.* [133, 135]. By permission of the publishers, Butterworth Heinemann Ltd. ©.

mation and crystal lattice parameters as high molecular weight PHB, although the melting temperature decreases with decreasing degree of polymerization.

When PHBV copolymers crystallize, an interesting phenomenon is observed. ^{13}C NMR spectroscopy [66] of these copolymers shows a Bernoullian distribution of comonomers at all compositions. Normally this would hinder crystallization of the polymer; however, after annealing, all the PHBV copolymers are nearly equally crystalline. This is possible

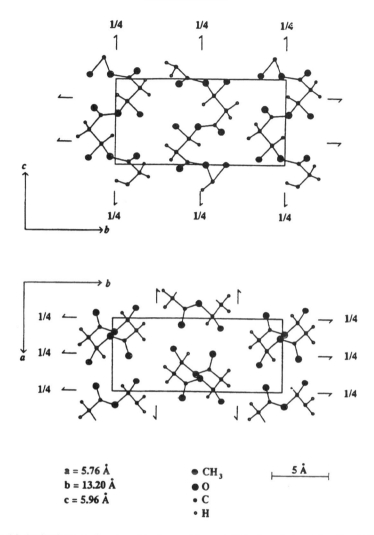

Figure 4.9 Projections on the *ab* and *bc* faces of the poly(β-hydroxybutyrate) unit cell. From Cornibert and Marchessault [101].

because of a phenomenon known as *isomorphism* [103], whereby two comonomer components cocrystallize in the same crystal lattice, such as that of the PHB homopolymer [66]. In the case of PHBV, two crystal lattices are involved: as the HV content increases, the crystal lattice suddenly changes to that of PHV. This phenomenon is known as *isodimorphism*. This behaviour is not seen with all copolymers; to exhibit this behaviour, the two comonomers must have similar shapes and volumes, and the chain conformations of both homopolymers must be compatible with either crystal lattice [104].

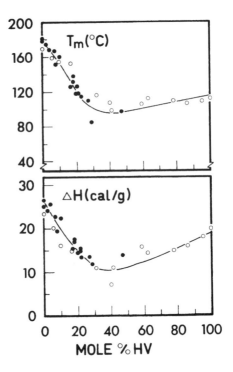

Figure 4.10 Variation of melting point (T_m) and enthalpy of fusion ($\triangle H_t$) with composition of bacterial (solid circles) and fractionated synthetic (open circles) poly(β-hydroxybutyrate). Redrawn and adapted, and reprinted with permission from Bloembergen *et al.* [105]. Copyright 1989 American Chemical Society.

As a result of the isodimorphic behaviour of PHBV, the melting point and enthalpy of fusion of the copolymers decrease with increasing HV content, pass through a minimum at ~ 40% HV, then increase to the values for PHV homopolymer. These effects were first demonstrated when values obtained from synthetic PHBV were included [105], to give results for all HV concentrations (Figure 4.10); only later was this achieved using biosynthetic material [108]. These results are expressed as depression of the PHB melting point and enthalpy of fusion caused by inclusion of HV units within the PHB lattice. Inevitably, the replacement of a CH_3 group by C_2H_5 results in lattice strain; when the lattice transforms to that of the PHV homopolymer, inclusion of HB units in the PHV lattice does not cause a similar effect. Correspondingly, there is a linear increase in the *a* crystal lattice parameter with increasing HV content up to 40% HV, followed by adoption of HV lattice dimensions [70]. More detailed studies of this behaviour have followed [105–113].

The phenomenon of isodimorphism has given the random copolymers from 0–25% HV composition (those which are currently commercially available) greater importance than would be expected. Normally, when crystallization proceeds with exclusion of the added unit, 10–20% of a different repeating unit causes a marked drop in the degree of crystallinity. In this case, the added unit is included in the lattice, providing a family of polymers with melting points varying from 180 to 100°C, which maintain a degree of crystallinity above 50% [66]. The lattice strain, due to cocrystallization of a repeating unit slightly larger than HB, is also responsible for enhanced toughness.

From a plastics technology perspective, control of the crystallization rate and thermal degradation are highly important. Because PHB and PHBV crystallize slowly, nucleation additives are used to promote the rapid formation of crystals throughout the sample. Fundamental studies on crystallization kinetics have provided reference data for PHB which indicate a nucleation rate which is many times slower than that of polyethylene [114–116, 138].

4.5.2 Nascent morphology

Another area of PHA research is that of nascent morphology, to determine how closely the polyesters in their 'as biosynthesized' state resemble the above models. As the commercial process for the isolation of PHBV is the isolation of the granules into a nearly pure suspension in water, understanding of the nascent state and its evolution as a function of the various process steps is economically important. For instance, PHBV can be isolated without drying, as a concentrated suspension of granules or latex product [35]; understanding the properties of this latex is required in order to optimize later processing [117].

Evidence for the presence of a membrane 2.5–4 nm thick surrounding PHB inclusion bodies has been reported for *Rhodospirillum rubrum* [118], *Bacillus cereus* [119, 120], *Ferrobacillus ferroxidans* [121], *Azotobacter chroococcum Beij* [87], *Bacillus thuringiensis* [122], *Pseudomonas oleovorans* [123] bacterial species and in the blue-green alga *Chlorogloea fritschii* [92]. Synthase and depolymerase activity appear to be related to this membrane, as several studies show loss of activity upon modification of the nascent morphology of the purified granules [90, 124–127]. For instance, chemical and physical treatments of the granules, such as sodium hypochlorite, acetone, trypsin, and freezing and thawing, damage the membrane and cause irreversible morphological transformation of the granule surface [125]. Further investigation is required to examine the nature of this membrane and its role in the enzymatic activity of the native PHB inclusion bodies [128].

The surface location of synthase enzymes implies appositional growth of PHA chains in the development of nascent morphology of the inclusions. Based on models of extended chain morphology of synthetic

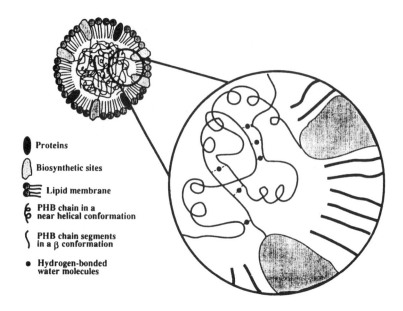

Proteins

Biosynthetic sites

Lipid membrane

PHB chain in a
near helical conformation

PHB chain segments
in a β conformation

Hydrogen-bonded
water molecules

Figure 4.11 Model representing the localization of the biosynthetic sites and other proteins in the membrane surrounding an inclusion body. The insert shows the different conformational arrangements that the polymer chains can adopt: near helical conformation, β conformation segments to facilitate hydrogen bonding to water, and random conformation. From Lauzier [133].

polymers [129–131], absence of entanglements is an important characteristic of inclusion nascent morphology. An attempt to exploit this characteristic to create extended chain fibres of PHA was partly successful [132]. Another important functional property is segment mobility, which allows enzymatic interaction within the inclusion and the exterior surface. Given the cyclic nature of PHA accumulation [61], depolymerase action is probably in juxtaposition with synthases at the inclusion surface. A recent model of PHA inclusions [133] is given in Figure 4.11; the basis of this model originates from the work of Ellar *et al.* [35], with important contributions from Dunlop and Robards [134].

The morphology of the PHA inclusions was originally suggested to be crystalline [35], just like that of the isolated dry granules. However, a recent high resolution ^{13}C NMR study showed that at least 70% of the molecules were mobile; thus the PHA inclusions must be mainly non-crystalline [75]. Scanning and transmission electron microscopy of freeze deformed PHB granules from a never-dried suspension isolated from *A. eutrophus* revealed that isolated granules are composed of a crystalline shell enclosing a core of non-crystalline material [135]. The solid shell was shown to be made of overlapping lamellar crystals, while the core under-

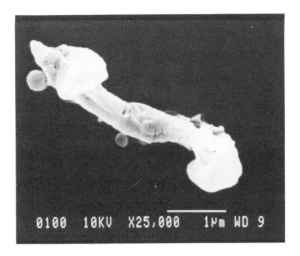

0100 10KU X25,000 1Pm WD 9

Figure 4.12 Scanning electron micrograph showing a highly deformed hypochlorite-isolated granule. The smeared core remains attached to the shell fragments of the granule. From Lauzier *et al.* [133, 135]. By permission of the publishers, Butterworth Heinemann Ltd. ©.

went ductile deformation at liquid nitrogen temperature (–196°C), as shown in Figure 4.12. These core molecules rapidly rearrange into shish kebabs or single crystal lamellae when deformed at 150–190°C, suggesting that they are in a metastable state in the nascent material.

It has been proposed [136] that water molecules form a pseudonetwork with the *in vivo* PHB molecules (Figure 4.13); upon isolation, the change of medium leads to topotactic crystallization at the surface of the granule, yielding the solid shell surrounding the non-crystalline core morphology. The lamellar crystals at the surface thus do not promote crystallization in the core, unless the granule is heated for the water to diffuse out. An alternative hypothesis for the lack of crystallinity in the core, based on a purely kinetic argument [116, 137], is that homogeneous nucleation is exceedingly improbable in PHB, hence crystallization would require millennia [138].

The lack of crystallinity in nascent granules results in significantly modified solubility characteristics, whereby never-dried nascent granules swell in acetone/water or methanol/chloroform mixtures and deposit lamellar single crystals upon drying [35]. This unusual morphology is in keeping with the model of a granule as shown in Figure 4.11. If the granules are lyophilized or spray dried, crystallinity is provoked. Lyophilization is physically stressful to the morphology [139], while spray drying is the equivalent of a heat treatment. The effect of various disruptive treatments (heat shock, ultrasound, etc.) and various drying processes (spray drying,

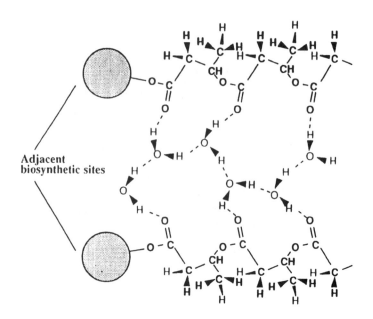

Figure 4.13 Model representing the hydrogen bonding of water molecules to two poly(β-hydroxybutyrate) chains from two adjacent sites in an early stage of the biosynthesis. From Lauzier *et al.* [133, 136].

freeze drying, etc.) on the granule morphology is still imperfectly understood.

4.6 Degradation

A major reason for the current interest in PHAs is their biodegradability. Having been made by bacteria, they can be completely degraded to carbon dioxide and water through bacterial action. The biodegradation processes are divided into two categories: intracellular and extracellular. Also of concern for those processing or using articles made of PHAs are non-biological degradative processes, such as thermal and hydrolytic stability. Numerous studies of the biodegradation of PHBV materials under various environmental conditions have been reported.

4.6.1 Intracellular biodegradation

Studies on degradation have been performed on *Bacillus megaterium* with extracts from *Rhodospirillum rubrum* [45, 90, 119, 125, 127, 140], and on *Alcaligenes eutrophus* [126, 141], *Zoogloea ramigera* [142, 143] and *Azotobacter beijerinckii* [48]. Generally, these processes are less well understood than the biosynthesis, but the best known intracellular degradation pathway can be represented as in Figure 4.14 [37, 48, 61], which combines with the biosynthetic pathway shown in Figure 4.4 to form the complete cycle of PHB metabolism. The PHB is depolymerized by a PHB-depolymerase enzyme, often in conjunction with a dimer hydrolase, to give R-β-hydroxybutyric acid. This acid is oxidized by a dehydrogenase enzyme to give acetylacetate, which is then esterified to acetoacetyl-CoA by an acetoacetyl-CoA-synthetase. This is degraded to acetyl-CoA by a retro-Claisen reaction catalysed by a β-ketothiolase enzyme; this enzyme is the only one used in both biosynthesis and biodegradation, and may be involved in regulation of the two pathways [49, 144]. In the presence of oxygen, the acetyl-CoA can then enter the tricarboxylic acid cycle (also known as the citric acid cycle) to be fully oxidized to carbon dioxide, with the corresponding release of energy.

Further investigation, however, is required to determine the finer details of this process. For instance, A. *eutrophus* is the only microorganism in which R-β-hydroxybutyric acid is the direct product of the depolymerisation by the depolymerase; in other microorganisms, the depolymerase degrades the PHB to dimers, which are subsequently cleaved by a dimer hydrolase. Depolymerase enzymes have been isolated from A. *eutrophus* [126, 145], B. *megaterium* [146] and R. *rubrum* [90], but their mode of action is not well understood; degradation of PHB native granules isolated from B. *megaterium* by exposing them to R. *rubrum* extracts was found to require three components: an esterase, a thermostable activator and a thermolabile depolymerase.

More is known about the action of the dimer hydrolase. This enzyme has been isolated from *Zoogloea ramigera* [147] and R. *rubrum* [148], and can hydrolyse not only the (R, R) dimer, but also, at a slower rate, (R, S) and (S, R) dimers (the latter by Z. *ramigera* only). The (S, S) dimer is not attacked by either enzyme. Higher oligomers (trimers, tetramers and pentamers) are attacked more rapidly than the dimer; for instance, the dimer hydrolase from R. *rubrum* hydrolyses trimer four times more rapidly than dimer [148]. Intracellular dimer hydrolases have also been isolated from *Pseudomonas lemoignei* [147, 149] and shown to be highly specific for the R configuration of β-hydroxybutyrate.

The cyclic nature of biosynthesis and degradation has been demonstrated for A. *eutrophus* [61, 150]. Provided that an excess of a carbon source is present in the bacterial culture medium, a limitation in nitrogen brings about an imbalance of growth which triggers the accumulation of PHB inclusions in the cells [16]. Although the accumulation of PHB is predominant under such conditions, degradation continues to take place, leading to a constant turnover of the PHB material within the inclusion bodies [150]. A separate study [60] showed that high molecular weight PHB was produced in the first stage of accumulation (up to ten hours),

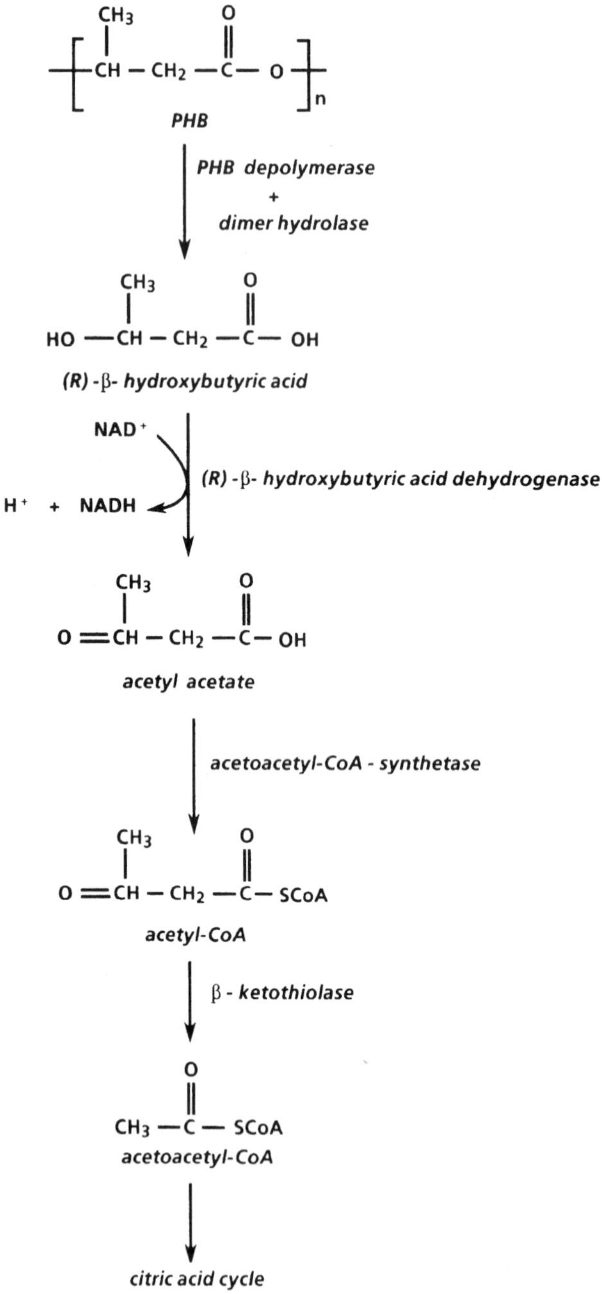

Figure 4.14 Intracellular biodegradation of poly(β-hydroxybutyrate).

then the molecular weight slowly decreased during the remaining period of accumulation. This result supports the cyclic nature of PHB accumulation, as well as implying high segment mobility of PHA inside the inclusions.

4.6.2 Extracellular biodegradation

By excreting depolymerase enzymes in the environment, some bacteria, like *P. lemoignei* [151, 152] or *Alcaligenes faecalis* T1 [153–156], can grow on extracellular PHA. The excreted depolymerases degrade the polymer into oligomers, mainly dimers. A dimer hydrolase is produced intracellularly [148] or is excreted (extracellular) along with the depolymerase enzyme [147, 149] to subsequently degrade these oligomers to monomer units. Depolymerase isolated from *Comamonas* sp. appears to have a different mechanism of PHB hydrolysis [157, 158], as the direct product of PHB hydrolysis by the purified enzyme is β-hydroxybutyrate.

The depolymerase produced by *P. lemoignei* has been shown [151] to be an exoenzyme, excreted in small amounts during the growth phase but mainly at its cessation. The presence of the product of hydrolysis, β-hydroxybutyric acid, also suppresses the enzyme excretion. The depolymerase has been purified [152] into four separate enzymes, each of which was able to hydrolyse PHB and its oligomers but was unable to degrade the dimeric ester. Analysis of the hydrolysis products led to the conclusion that these depolymerases have affinity with the hydroxyl rather than the carboxyl end of the oligomers, and that cleavage of the ester groups occurred at every second or third bond from the hydroxyl terminus [152].

A depolymerase enzyme isolated and purified from *A. faecalis* T1 [153, 159] showed strict specificity towards PHB polymer that had been chemically purified but not towards the native granules. Analysis of the hydrolysis products of *R*-β-hydroxybutyrate labelled at the hydroxl terminus showed that [154] the enzyme cleaves only the second ester linkage from the hydroxyl terminus of the trimer and tetramer and acts as an endo-type enzyme toward the pentamer and higher oligomers. Its affinity towards PHB, an amphiphilic substrate [117], is thought to be due to the presence of a hydrophobic site on the enzyme which could interact with the polymer, in keeping with recent concepts from lipase and cellulose structure studies [160, 161].

The purified *A. faecalis* T1 depolymerase enzyme was used in three separate studies to monitor the degradation of solution cast films of PHB [162, 163] and of PHBV [164]. The results of the first study [164] showed that surface degradation of the films was apparent without significant weight loss and that the hydrolysing activity of the enzyme was greater for PHB than for PHBV. This suggests that *A. faecalis* depolymerase first hydrolyses chains in the non-crystalline regions, then proceeds to depolymerize chains in the crystalline regions.

4.6.3   Thermal degradation

Understanding of the thermal degradation of PHAs is required so that processing can be designed to minimize structural damage. The mechanism of thermal degradation of PHAs is well understood in terms of the McLafferty rearrangement; in fact, PHBV is so susceptible to thermal breakdown that pyrolysis yields of crotonic acid approach 90% [165]. To control this aspect of degradation during processing, suitable additives are used; however, caution is required if prolonged heat exposure of PHBV moulded articles is proposed. Early studies on PHB thermal degradation were reported [165] by Morikawa and Marchessault and by Hauttecoeur *et al.* [166]. These two studies emphasize controlled degradation to obtain oligomers with terminal double bonds instead of hydroxyls.

A detailed study of thermal degradation of PHB identified the volatile products of degradation [167]. When heated from 0 to 338°C under vacuum, PHB releases isocrotonic acid (0.9 wt%), crotonic acid (35.3 wt%) and the dimer (41.2 wt%), trimer (12.5 wt%) and tetramer (2.9 wt%) of PHB. When the heating is continued to 500°C, traces (4 wt%) of the degradation products of these volatiles are observed: carbon dioxide, propene, ketene, acetaldehyde and β-butyrolactone. Measurement [168] of the molecular weight loss of PHB during isothermal degradation at 170–200°C shows that thermal degradation occurs through a random scission process involving a six-membered ring ester decomposition similar to the Chugaev reaction [169]. This reaction leads to an increase in molecular weights early in the degradation process [170].

4.6.4   Hydrolytic degradation

As PHB-utilizing bacteria and fungi are not present in the human body, degradation of devices *in vivo* must proceed via other routes, such as hydrolysis. This may be catalysed by enzymes from the body's immune system [3]. Hydrolytic degradation of PHB and PHBV has been studied both *in vivo* and *in vitro* [171–174]. *In vitro* studies of the degradation rate of solution and melt-cast films of PHBV in a phosphate buffered medium [171] showed an increase in degradation rate with increasing HV content, or with increasing temperature. Prior melt processing significantly affected the rate of hydrolytic degradation [174]. Although sample weight and tensile strength remained constant for several months, surface modification and an increase in porosity of the matrix occurring during this time allow for later accelerated degradation [173]. *In vivo* degradation studies of PHB monofilaments [172] showed that for degradation to occur, they had to be predegraded with 10 Mrad γ-irradiation. Increased HV content in the PHBV copolymers did not increase the rate of *in vivo* degradation, and even retarded the rate of high temperature hydrolysis.

Studies of the heterogeneous degradation of solution-cast films of PHB and PHBV (68 mol% HV) in a buffered medium [175] confirmed that a random chain scission hydrolytic degradation process was taking place in both the crystalline and the non-crystalline domains. A two-stage hydrolytic process was suggested: random scission of the chains leads to a decrease in molecular weight, triggering weight loss of the films below a molecular weight of 13 000. An apparent increase in crystallinity with hydrolysis time was explained as due to crystallization of chain fragments in hydrolysed amorphous regions; an alternative explanation is that preferential hydrolysis of chains in the amorphous regions leaves an increasing proportion of the chains in the crystalline zones [176]. A later hydrolytic degradation study on PHBV (14 mol% HV) fibres [177] showed that the tensile strength of the fibres only started to decrease when the molecular weight fell to 70 000, reaching a zero value by the time the molecular weight was 17 000; the fibre weight decreased only after the molecular weight was below 17 000.

Recent studies on the hydrolysis of PHB by 3N HCl at 104.5°C [133] have shown that the acid attacks the ester linkages of both the crystalline shell and the non-crystalline core. During the first six hours, weight loss was moderate and the crystallinity increased steadily; after this time, the weight loss rate constant nearly doubled and crystalline perfection decreased. Of note in this study was the retention of the granule texture after extensive chain cleavage, and the high crystallinity of the residue after hydrolysis.

4.6.5 Environmental degradation

Of greatest relevance to the user of articles made of PHB or PHBV are the rates of degradation of these articles under various environmental conditions. Biodegradation normally proceeds via surface attack by bacteria, fungi and algae [29, 178]. These microorganisms can excrete extracellular enzymes to solubilize the PHB surface on which they are growing. The soluble degradation products are then absorbed through the cell wall and metabolized. Because of the importance of surface attack, rates of biodegradation depend in part on the ease of surface colonization. Surface area and thickness are important, as is surface texture in environments where nascent bacterial colonies might be physically washed off smooth surfaces.

The actual degradation times depend on the environment involved, as well as on the thickness of the article. These factors control biological oxygen demand and supply of nutrients essential to microbial growth, such as nitrogen and phosphorus. Studies performed in various environments [29] show the PHB degradation rate progressively decreasing in anaerobic sewage, well watered soil, sea water sediments, aerobic sewage, the rumen

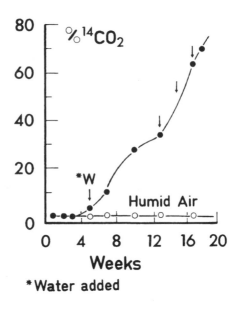

Figure 4.15 Biodegradation of ¹⁴C-tagged poly(β-hydroxybutyrate-co-β-hydroxyvalerate) in moist soil. Redrawn and adapted from ICI Americas [180].

of cattle and sea water [179]. Slower degradation rates were seen *in vivo* (subcutaneous or intramuscular) and *in vitro*.

An example of the rates of biodegradation is shown in Figure 4.15 [180] for ¹⁴C-tagged PHBV film buried in moist soil; approximately 50% of the carbon content was lost as CO_2 within 16 weeks. A 1 mm thick moulding resin test bar is reported [181, 182] to break down completely within six weeks in anaerobic sewage; this time increases to 60–70 weeks in aerobic sewage and soil at 25°C, and to 350 weeks in sea water at 15°C. Packaging films 50 μm thick totally degrade in 1–2 weeks in anaerobic sewage [183], 7 weeks in aerobic sewage, 10 weeks in soil at 25°C and 15 weeks in sea water at 15°C. A molded shampoo bottle made of PHBV is almost completely degraded after 15 weeks in a compost pile [184]; significant degradation in a landfill occurs over 40 weeks.

Given that PHBV degrades under such a variety of conditions, one concern could be its shelf life. Fortunately, the rate of PHB degradation in moist air is negligible [29], giving it an acceptable shelf life. More serious is that a lack of standardized testing procedures for the evaluation of biodegradation [185] makes it difficult to compare results by different researchers, working under different conditions on different types of samples. This limitation may soon be resolved, as PHBV is being used as one of the standard reference materials for correlating lab-to-full-scale tests [186, 187]. Thus, other polyester-type degradable materials may soon be calibrated by comparison with PHBV reference samples. Preliminary results indicate that among biodegradable thermoplastic candidate materials, only poly(ε-caprolactone) degrades at the same rate as PHBV [188].

4.6.6   *Effects on recycling*

As the current interest in PHB is largely based on its biodegradability, to consider large scale recycling of this polymer is to ignore one of its most important properties. However, it is useful to consider the potential effects of PHB as a contaminant in the recycling stream of conventional synthetic polymers, such as polyolefins. The thermal treatments will almost certainly cause thermal and mechanical effects with significant molecular weight loss in the PHB. This could produce free crotonic acid from the hydroxybutyrate and double bonds at the end of the PHB chains. Possible crosslinking could result due to the vinylic character of this system. If the quantity of PHB is not too great, then this should not cause any significant problem in a mixed recycling stream. For large quantities of PHB, conventional recycling methodology used for synthetic polymers is probably not applicable; in this case, a composting or managed landfill disposal system would be more appropriate.

4.7   Applications

A variety of applications for PHB and PHBV can be envisaged, thanks to the wide range of specialized properties of these materials. Although the first interest in these materials was due to their biosynthesis from renewable resources and thus independence from oil supplies, most of the current interest relates to their biodegradability. This permits waste reduction, as well as the creation of value-added materials where degradation is the functional characteristic of the product. The chirality of the polymer also allows use in more upstream applications, such as chiral blocks for organic synthesis.

One of the simplest applications for PHBV is as a biodegradable substitute for polyolefin containers, plastic films and bags [189]. The first commercial use of PHBV is along these lines: Wella AG, Darmstadt (Germany) is using an injection blow moulded bottle to package a biodegradable hair shampoo [190]. PHBV has also been used for motor oil containers and disposable razor handles [191]. Such applications of PHBV are particularly an asset for items difficult to separate for recycling, such as kitchen films, diapers and sanitary napkins [29].

The gas barrier properties of PHBV could lead to applications in food packaging, or as a replacement for poly(ethylene terephthalate) for plastic beverage bottles [96]. These gas barrier properties also suit PHBV for use in coated paper and films, such as coated paper milk cartons [190]. The polyethylene coating currently used for this purpose is non-biodegradable, preventing either biodegradation of the paper or routine recycling of the component fibres; PHBV coated paper has been shown to be completely biodegradable and also easier to recycle than conventionally coated paper [117, 192].

An unusual powder deposition technique, direct electrostatic coating (DEC), has been used for depositing PHB on a low dielectric substrate such as paper [193]. The electrostatically adhering powder is subsequently fused and calendared to yield a clear barrier film [194]. The adhesion of this layer is immensely superior to that of melt-laminated film. Powder deposition and biodegradation are the functional properties behind a recent US patent [195] for use of PHBV as a toner in xerography, the difficulties of detaching plastic inks being facilitated by chemical and enzymatic methodology, during paper recycling operations.

More specialized applications of PHBV include controlled release [196], where biodegradation is required while the item is in use. As *Alcaligenes eutrophus*, the organism used for commercial production of PHBV, was first isolated from soil, PHBV is naturally suited to agricultural applications requiring biodegradation in the soil. For example, insecticides incorporated into PHBV pellets and sown with the farmer's crop would be released at a variable rate related to the level of pest activity, as the bacteria degrading the polymer would be affected by the same environmental factors as soil pests [29]. Similarly, as PHBV degrades well in the rumen of cattle, it can be used as a biodegradable matrix for drug release in veterinary medicine. Boluses containing medicine can remain in the rumen to deliver doses of the medicine at a preset time interval [197].

PHBV can also be used internally in humans, as the polymer itself is non-toxic and compatible with living tissue, and the sole degradation product is R-β-hydroxybutyric acid, which is a normal mammalian metabolite found in concentrations of 3–10 mg/100 ml of blood in healthy adult humans [198]. Low molecular weight PHB has also been detected, bound primarily to albumin but also to low density lipoproteins, in human blood serum [42]. Studies of obese patients undergoing therapeutic starvation have shown that sodium β-hydroxybutyrate can be used as an intravenous or oral carbon supply [199]. Relative to the more common glucose drip, body protein conservation is improved, without reducing the rate of weight loss. An acyl derivative of the polymer has been patented for use as a food emulsifier or shortening aid [200].

For controlled drug release applications in human medicine, PHBV is used not as a bolus, but as microcapsules [201], which are injected subcutaneously as a suspension [202] or pressed into a pill and administered orally [203–206]. As a medical material, the most notable features of PHBV are that it is very biocompatible, producing an extremely mild foreign body response, and that the biodegradation rate *in vivo* is slow. For example, a monofilament surgical suture would require several years to be totally resorbed by the body; as the time taken for biodegradation is related to surface area, multifilament sutures or microcapsules resorb much more quickly.

Typical applications of PHBV in hospitals would be as surgical swabs, wound dressings and lubricating powders for surgeons' gloves [207]. A blood-compatible membrane has also been proposed [208]. ICI has developed a centrifugal spinning process to make cotton wool-like products and gauzes from concentrated PHBV solution [198]; after treatment with biocompatible surfactants, the intrinsically hydrophobic fibres are much like absorbent cotton wool [29]. However, unlike cotton fibres, bits of PHBV fibre from the swab or dressing can be left in the wound without concern, as they will biodegrade.

A high technology futuristic use of PHBV could be as a vascular graft or blood vessel, composed of very fine fibres arranged to form a water-impermeable tube of suitable internal diameter [29]. This could act as a temporary scaffold for new tissue in-growth, eventually being completely replaced by natural tissue. This avoids the problem of thrombus formation and eventual blockage in synthetic arteries, which arises as a direct result of the body's response to the non-degrading foreign matter of the vessel wall. Surgical implants of PHB and PHBV to join tubular body parts have already been developed [209], as have sheets [210] or coils [211] of PHBV to separate tissue in wound healing.

Other applications of PHBV may be possible due to its piezoelectric response [212]. As the size of this effect is an order of magnitude smaller than that of poly(vinylidene fluoride) (PVDF), it is unlikely that PHBV will replace oriented PVDF in hi-fi or other electronic equipment [29, 30]; however, PHBV may be useful where a temperature-independent response is required. For instance [29], PHBV may be suitable to use in a pressure sensor in a heat-sealing device, where PVDF would not be suitable because its pyroelectric response would interfere with the strain piezoelectric effect.

In addition, the piezoelectric properties of PHBV are similar to those of natural bone. It is known that bone can be strengthened and repaired by electrical stimulation [213]; therefore, a bone fracture fixation plate, made from a reinforced PHB composite to match mechanical properties, might actually stimulate bone growth and healing. An added advantage of such a plate is that it would be biodegradable, and could be left in place to be slowly resorbed by the body, instead of requiring a second operation to remove it.

Further applications of PHB may arise from its optical activity and synthetic potential. Chromatography is potentially one such application, as a chiral stationary phase can be used in chromatography to separate optical isomers [214]. PHB can also be hydrolysed to provide optically pure monomer for use as a chiral building block in organic synthesis [29, 37, 215]; compounds already reported include the fungicides norpyrenophorin, pyrenophorin and vermiculine [216], the macrocyclic component of the antibiotic elaiophylin [217], the (S, S) enantiomer of the natural product grahamimycin A_1 [218] and pheromones of the smaller European elm bark beetle [219] and the western corn rootworm [220]. The biosynthesis of PHBV also forms the basis of a process to denitrify drinking water [221–223].

Commercially, PHBV is available as spray dried powder (Figure 4.7) in the range of 0–25% HV [180]. The concentration of HV units affects both the properties and the rate of degradation of the materials, making certain applications most feasible at certain HV concentrations. Pure PHB (0% HV) is considered useful for applications where the brittleness is not a problem, and purity of the monomer is required: these include medical implants, pharmaceutical applications, and use as chiral building blocks. A slight reduction in brittleness (5% HV) makes the material suitable for relatively stiff injection mouldings, while copolymer containing 10% HV is useful for injection moulding, extrusion and injection blow moulding for packaging materials. Finally, PHBV with 20% HV is suited to slow release veterinary and medical applications.

Different grades of PHBV are also available, depending on the anticipated use [180]. Most economical are the technical or standard grades with 95–96% purity and molecular weights of 400 000–750 000. These can be processed without additives, although for most melt processing routes, use of a nucleating agent, such as boron nitride, talc, micronized mica or chalk, is recommended. If desired for property modification, plasticisers, fillers or pigments can be melt compounded with PHBV. For medical and pharmaceutical uses, specially purified grades with 99.5% purity are available in the molecular weight range of 30 000–750 000. Filled grades containing hydroxyapatite are also available for medical uses. If stored under clean, dry conditions, no preparatory treatment of the powder or granules should be required before processing.

4.8 Economics

The main factor behind the emphasis on specialty rather than high volume applications for PHBV is cost. At current production levels (660 000 lb/year), the price is US$8–10/lb [187, 191]; this is projected to fall to US$4/lb by 1996, when a new plant capable of producing 10–20 million lb/year begins manufacture. This price makes it difficult to compete with bulk polypropylene at US$0.30–0.45/lb [224]. Numerous examples exist of technological improvements together with the economies of scale dramatically reducing the cost of new materials [32]; however, with current prices, small, high-value-added, specialty applications, rather than commodity-scale production and commodity plastic applications, are the likely outlets for PHBV.

The cost of PHBV production has three main components: materials, mainly the carbon source; the fermentation process itself, including polymer isolation and purification; and capital-related charges [29]. Improvements in fermentation science and polymer extraction technology, as well as economies of scale, have improved the competitiveness of the biopolyester; now the high cost is due mainly to the cost of substrate [67] and the separation process. Alternative lower cost substrates include materials such as methanol [225], molasses [226, 227] and hemicellulose hydrolysate [228]. The substrate cost per metric tonne of PHB produced has been calculated for a variety of substrates, taking into account the variation in PHB yield with substrate (Table 4.3) [229]. Clearly, replacing glucose as the substrate could greatly reduce the cost; for instance, use of lactose in place of glucose could reduce substrate costs by 86%. Even

Table 4.3 Effect of substrate cost per metric tonne (m.t.) on total production cost of PHB*

Substrate	Approximate cost ($US/m.t. of fermentable substrate)	Yield of PHB/ substrate[6]	Substrate cost/PHB ($US/m.t. PHB)
Methanol	184[1]	0.18	1020
Ethanol	502[1]	0.50	1005
Acetic acid	705[1]	0.33	2135
Glucose	493[2]	0.33	1495
Fructose	517[2]	0.33	1565
Sucrose	790[1]	0.33	2395
Cane molasses	220[3]	0.42	524
Lactose (cheese whey)	71[4]	0.33	215
Hemicellulose hydrolysate	69[5]	0.20	345
Lactic acid (fermented cheese whey)	173[3]	0.33	525

[1] Prices from Chemical Week, Chemical Marketing Reporter.
[2] Casco Inc., Pointe Claire.
[3] Canada West Indies Molasses Company Ltd., Montreal, Quebec, Canada.
[4] Saputo Fromage Ltd., St. Leonard, Quebec, Canada.
[5] Based on economic estimation of Lynd (1989), acid pretreatment followed by enzymatic hydrolysis. Assumes zero cost for waste hemicellulose.
[6] All yields taken from Byrom (1987) except those of cheese whey, fermented cheese whey and hemicellulose hydrolysate which were estimated.

Byrom, D. (1987) *Trends Biotechnol.*, **5**, 246–250.
Lynd, L. R. (1989) *Adv. Biochem. Eng. Biotechnol.* **38**, 1–52.

* The authors are grateful to Dr. Juliana A. Ramsay, Ecole Polytechnique in Montreal, for providing the information in this table.

more economical in terms of substrate cost would be the use of photosynthetic bacteria such as *Rhodobacter* or *Rhodospirillum* to produce PHB
from sunlight [190]; this is only practical if substantial improvements in
productivity can be achieved.

4.9 Future prospects

PHAs are under active investigation by many researchers, and new members of this family are still being obtained. More detailed examinations of
homopolymer properties and a wider range of structures can be designed
using chemical synthesis. Blending PHAs with other polymers allows
further modification of properties and adjustment of the rate of degradation, as well as reduction in cost. Progress is also being made in transgen-
ics, which may someday allow production of PHAs in plants such as
tobacco or rapeseed. The latter produces lipids in an amount of 40% by
weight, and the biosynthesis mechanisms of PHB and lipids are almost
identical.

4.9.1   Other bacterial PHAs

PHB and PHV are only the first two members of a family of biopolyesters
of the general structure shown in Figure 4.16. Bacterial PHA with long
side-chains (LSC) were first isolated by Wallen and Rohwedder [62] from
estuarian sediments; in 1983, De Smet *et al.* [72] discovered that *Pseudomonas oleovorans* bacterial cultures could be force fed an *n*-octane substrate to produce terpolymers consisting chiefly of β-hydroxyoctanoate.
This led to a growing number of biochemical studies, in which various
feed sources were supplied to a wide variety of bacteria in order to
obtain a range of polyesters with different side-chain lengths. Most LSC
polyalkanoates reported are produced either by *Pseudomonas oleovorans*
or *Pseudomonas putida* [72, 73, 230–235]; long side-chain PHAs with up
to nine carbons in the side-chain have been reported from *Pseudomonas
oleovorans* fed on *n*-alkanes and alkanoic acids [74, 236]. However, copolymers of PHB and poly(3-hydroxy-2-butenoate) can be produced by a
Nocardia [76] species grown on butane, and use of *Bacillus megaterium*
[64] has also been reported. Bacteria which accumulate LSC PHA do not
produce PHB under standard microbiological taxonomic tests [190].
  The properties of various LSC PHAs are reported in Table 4.4. Unlike
the thermoplastic materials PHB and PHBV, the LSC PHAs are thermoplastic elastomers, with melting points in the 45–60°C range and T_gs down
to 40°C (Figure 4.17). The LSC PHAs are not homopolymers, but are
named after their major component for reasons of simplicity. For instance,
PHO produced by *P. oleovorans* [237] may contain up to 85 mol% HO

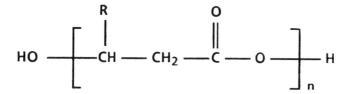

R =   methyl;   β-hydroxybutyrate *(HB)*

R =   ethyl;   β-hydroxyvalerate *(HV)*

R =   propyl;   β-hydroxycaproate *(HC)*

R =   butyl;   β-hydroxyheptanoate *(HH)*

R =   pentyl;   β-hydroxyoctanoate *(HO)*

R =   hexyl;   β-hydroxynonanoate *(HN)*

R =   heptyl;   β-hydroxydecanoate *(HD)*

R =   octyl;   β-hydroxyundecanoate *(HUD)*

Figure 4.16 Generalized chemical structure of poly(β-hydroxyalkanoates).

Table 4.4 Physical properties of poly(3-hydroxyalkanoates)[a]

Sample[b]	Average monomer mass (g/mole)	Average[c] side-chain length (C atoms)	T_g onset (°C)	T_m (°C)	Density (g ml^{-1})
PHB	86.09	1.00	5	180	1.2430
PHV	100.12	2.00	−11	105–108	1.20
PHC	121.16	3.50	−25	N/A	N/A
81.5 mol% HC					
PHH	128.17	4.00	−33	45	N/A
93.5 mol% HH					
PHO	141.08	4.99	−36	61	1.019
86.1 mol% HO					
PHN	146.69	5.32	−39	54	1.026
58.4 mol% HN					
PHD	153.00	6.25	−40	54	1.033
11.1 mol% HD					

[a] Data obtained from references 74, 99, 238; [b] PHB: poly(β-hydroxybutyrate), PHV: poly-(β-hydroxyvalerate), PHC: poly(β-hydroxycaproate), PHH: poly(β-hydroxyheptanoate), PHO: poly(β-hydroxyoctanoate), PHN: poly(β-hydroxynonanoate), and PHD: poly(β-hydroxydecanoate); [c] Derived from true compositional data from reference [74].

N/A = Not available

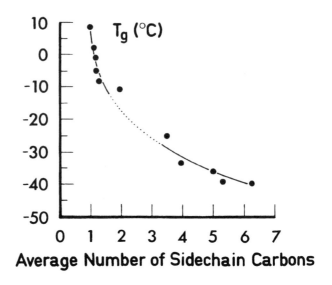

Figure 4.17 Glass transition temperature (T_g) of poly(β-hydroxyalkanoates) as a function of the number of carbons in the side-chain. Redrawn and adapted from Marchessault *et al.* [238]. By permission of the publishers, Butterworth Heinemann Ltd. ©.

units, probably randomly mixed with other repeating units, such as β-hydroxycaproate (HC) and β-hydroxydecanoate (HD). The predominant repeating unit in LSC PHAs has the same number of carbons as the substrate, while the other repeating units differ by plus or minus two carbons due to the β-oxidation process of bacterial physiology.

PHAs with long side-chain alkyl groups between pentyl and heptyl are proposed to have a minimum energy conformation with a two-fold screw axis and to crystallize in an orthorhombic or monoclinic lattice with two molecules per unit cell, just like PHB and PHBV [238]. The 4.55 Å helical pitch is much lower than the values of 5.96 Å and 5.56 Å for PHB and PHV, respectively, as a result of the increased side-chain interaction [238]. These polymers crystallize with alkyl side-chains in an extended conformation to form ordered sheets in one of two possible models [238], as shown schematically in Figure 4.18.

Polyesters with functionalized side-chains, attractive for novel drug delivery systems [182], have been produced by bacteria upon modifying the feed substrate. Poly(β-hydroxyalkenoates), which contain unsaturation in the side-chain, can be produced by feeding *Pseudomonas oleovorans* on 1-alkenes or alkenoic acids [230, 233, 236]. These materials, like the saturated LSC PHA, are also of low crystallinity compared to PHB and

(a)   (b)

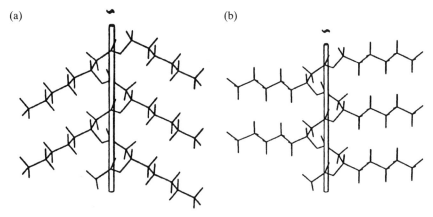

Figure 4.18 Two models of the poly(β-hydroxyoctanoate) 2_1 helix: (a) herringbone; (b) double comb. From Marchessault *et al.* [238]. By permission of the publishers, Butterworth Heinemann Ltd. ©.

PHBV, and are thermal elastomers rather than thermoplastics. The degree of unsaturation can be varied from 0 to approx. 50% by using various ratios of *n*-alkanes and 1-alkenes as substrate. When 1-octene and 1-decene are used as the carbon source, the stored polyester consists of both β-hydroxyalkanoate and terminally unsaturated β-hydroxyalkenoate, of which the pendant group varies in length from propyl to heptyl. PHAs have also been prepared with methyl-branched five-carbon side-chains [234], as have PHAs with brominated [239] and cyano-containing [240] alkane side-chains.

Recently, copolyesters of 3-hydroxybutyrate (β-hydroxybutyrate) and 4-hydroxybutyrate (γ-hydroxybutyrate), P(3HB-co-4HB), have been obtained from *Alcaligenes eutrophus* fed on 4-hydroxybutyric and 4-chlorobutyric acids [241] or γ-butyrolactone [242, 243] (Figure 4.19). These copolyesters can be prepared with a range of comonomer compositions similar to PHBV, and have a random sequence distribution of monomers [241]. The polymers range from crystalline to elastomeric, depending on composition; incorporation of up to 27% 4HB units appreciably enhances the rate of enzymatic biodegradation relative to PHB or PHBV [164, 244]. Increasing the 4HB content from 0 to 85 mol% in solvent-cast films crystallized at room temperature for two weeks results in a drop in crystal-

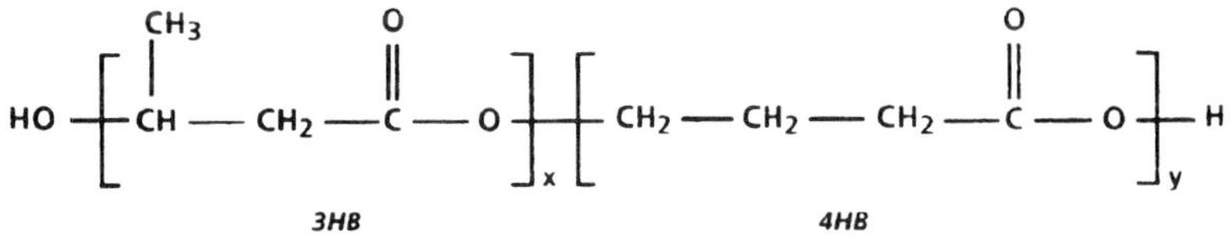

$$HO \left[\begin{array}{c} CH_3 \\ | \\ CH \end{array} - CH_2 - \begin{array}{c} O \\ \| \\ C \end{array} - O \right]_x \left[CH_2 - CH_2 - CH_2 - \begin{array}{c} O \\ \| \\ C \end{array} - O \right]_y H$$

                 3HB   **4HB**

Figure 4.19 Chemical structure of poly(3-hydroxybutyrate-co-4-hydroxybutyrate).

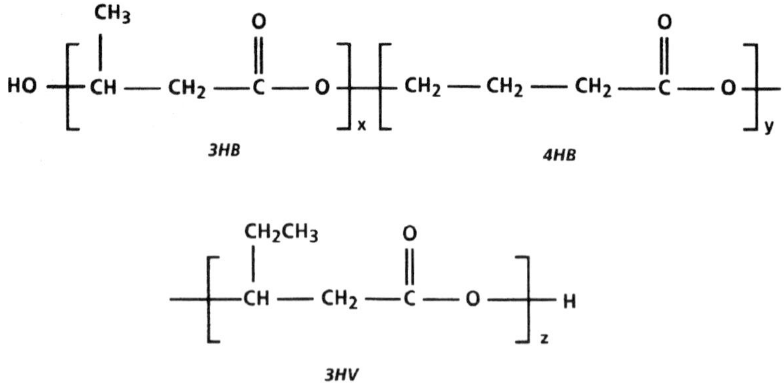

Figure 4.20 Chemical structure of poly(3-hydroxybutyrate-co-4-hydroxybutyrate-co-3-hydroxyvalerate).

linity from 59 to 29%, and a drop in melting point from 180 to 48°C [244]. P(3HB-co-94%4HB) was noted to have an almost identical crystallization rate to that of P(3HB) homopolymer, but with a unit cell corresponding to poly(4-HB). This copolymer system does not display the isodimorphism phenomenon [244] because the crystalline conformational features of 4-HB repeating units are different from those of 3-HB.

In addition, P(4HB) has been biosynthesized by *Alcaligenes eutrophus* using a substrate of 4-hydroxybutyric acid, citrate and ammonium sulphate [244]. A new terpolyester (Figure 4.20) containing 3HB, 4HB and 3HV units has been produced in *Alcaligenes eutrophus* grown in a culture medium containing 4-hydroxybutyric acid and pentanoic acid [245]. The 4HB fraction in the terpolyesters increases with increasing fraction of 4-hydroxybutyric acid, while the 3HV fraction increases by increasing the ratio of pentanoic acid. Another copolyester of 3-hydroxybutyrate and up to 7% 3-hydroxypropionate was reported in 1991 [246]; this also has a higher susceptibility toward enzymatic biodegradation than bacterial PHB.

Of the wide range of polyalkanoates isolated in the last decade, most are based on β-substituted alkanoates. The occurrence of γ-substituted alkanoates such as P(4-HB) is remarkable because ring-opening polymerization of γ-lactones is virtually unknown [247]. To date, the only synthetic polyalkanoates with properties comparable to this family to receive serious commercial consideration are polypivalolactone [248] and poly(ε-caprolactone). The latter is in commercial production and demonstrates good biodegradabilty characteristics.

4.9.2 Synthetic PHAs

Numerous studies involve the preparation of synthetic PHAs, usually from the ring-opening of β-lactones. Presently, this is not commercially viable on a production scale in comparison to bacterial fermentation, due to the expense of preparing the lactone monomers and, more particularly, the polymerization catalyst. Aluminum alkyl catalysts, which produce synthetic polyesters most resembling those of bacterial origin, yield only 10–20% stereoregular isotactic polymer from starting monomer [249]; use of a preformed isobutylaluminoxane catalyst improves the yield to 37% [250]. The immediate significance of synthetic PHAs thus lies not as a production scale thermoplastic polyester, but as a model to better understand the properties of the biosynthesized materials.

An example of synthetic PHAs being used to aid in the understanding of the biosynthesized materials was the preparation of synthetic PHBV with a full range of composition, from 0–100% [105], in order to demonstrate the phenomenon of isodimorphism in PHBV; at the time, biosynthesized samples were available only up to 47% HV. Another advantage of synthetic polyesters is that homopolymers can be produced; most bacterial polymers, especially those containing long alkyl side-chains, are usually terpolyesters [74]. Thus, physical and chemical properties specific to the bacterial LSC homopolymer are presently measurable only from the synthetic polyesters [286, 287]. These measurements are of significance in predicting structure-property relationships necessary for designing materials for specific applications.

The potential of synthetic PHAs also lies in the possibility of being able to tailor a polyester with substituents at the β-carbon centre not available from biological sources. An example is the synthetic copolyester of β-hydroxybutyrate and β-benzyl malate (R = $COOCH_2C_6H_6$ in Figure 4.16) [250]. This material has not been reported to have a bacterial analogue and is a potential biodegradable material, as it can undergo debenzylation to yield a hydroxybutyrate–malic acid copolymer.

Another feature of synthetic PHAs is that stereocopolymers can be made by using racemic and optically active lactone monomer. For example, atactic PHB is easily prepared from racemic β-butyrolactone and an organozinc catalyst [251], while isotactic analogues of the bacterial polymer can be made using organozinc and organoaluminum catalysts with optically pure β-butyrolactone [252]. Isotactic polymers containing blocks of R and S sequences along the chain can also be made from racemic β-butyrolactone and organoaluminum catalysts [253]. Recently, syndiotactic PHB, with alternating R and S centres, has been prepared [254, 255]. These stereoblock copolymers undergo biodegradation at rates which are comparable to that of the biosynthesized polymer [256, 257]. Furthermore, the rate of biodegradation is a function of polymer tacticity, with intermediate tacticity samples being much more biodegradable than higher isotacticity ones. Figure 4.21 shows the results of heterogeneous hydrolysis of synthetic PHB by an extracellular depolymerase from *A. faecalis* [257]. The favourable conjunction of low crystallinity and high chain mobility in water is a

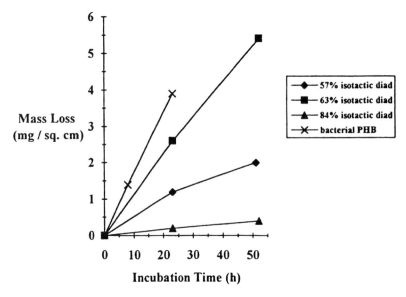

Figure 4.21 Enzymatic degradation of bacterial (100% isotactic) and synthetic poly (β-hydroxybutyrate) as a function of tacticity. Extracellular depoymerase from *Alcaligenes faecalis* was used. From Jesudason *et al.* [257].

proposed explanation for why intermediate tacticity samples are so readily degraded by enzymes [257].

4.9.3 Blends

Another approach to modifying the properties of PHB and PHBV, which also reduces the cost of the final material, is to form blends of the biopolyester with other materials. Studies have shown miscibility between PHB and poly(ethylene oxide) [258], poly(vinyl acetate) [259], poly(vinylidene fluoride) [260] and poly(vinyl chloride) [261, 262]. Completely biodegradable blends of PHA and starch have also been investigated [263–265]. The presence of starch was found to increase the rate of PHA degradation, eventually resulting in 100% degradation; to maintain acceptable mechanical properties, the concentration of the starch component must remain below about 40%.

Blends have also been formed between different PHAs, or between different stereoisomers of PHB. These approaches are more likely to achieve good miscibility. Thus, bacterial PHB has been solution blended with PHBV (8% HV), leading to crystalline compatibility (co-crystallization) shown by the presence of a single melting endotherm [266]. Blends of atactic synthetic PHB with bacterial PHB significantly improve the susceptibility of the biopolymer to enzymatic degradation, with a blend of 50 wt% showing the highest rate of degradation [267]. These solvent-cast films are also tougher and less stiff than those of the PHB bacterial homopolymer. When a partially isotactic synthetic PHB sample was blended with bacterial PHB, a single melting point lying between those of the pure components was observed at all compositions [268]. This is evidence for the miscibility of the crystalline phases of these materials due to isomorphic crystallization.

Melt miscibility of PHBV with commercial thermoplastics is under active investigation. Of particular interest is the case when the commercial thermoplastic itself is biodegradable. In studies where PHB [269] or PHBV [270, 271] was melt-blended with cellulose acetate butyrate (CAB) or cellulose acetate propionate (CAP), blends up to 50% PHB or PHBV were found to be miscible and amorphous, while blends with over 50% PHB or PHBV were semi-miscible and semi-crystalline. PHBV in these blends was found to crystallize in a different morphology to pure PHBV, causing these blends to have different physical properties, such as better tear strength, than either of the pure components [270]. Morphologically different situations could also be achieved by heat treating these blends, which may allow adjustment of morphology to suit biodegradability requirements [269].

4.9.4 Genetic engineering

An exciting development in PHA synthesis is the cloning of the genes in *A. eutrophus* that tell the bacterium how to make PHB [189, 272–277]. These genes have been transferred into *Escherichia coli*, the standard laboratory bacterium, allowing all the tools developed for working on *E. coli* to be put into use on the PHB genes. For instance, one *E. coli* mutant has cells that are ten times larger than normal; putting the PHB genes into this form of *E. coli* might produce a bacterium that makes ten times as much polymer per cell, making the polymer easier to produce. Polymer isolation can also be simplified by inserting the PHB genes into a strain of *E. coli* which bursts when heated to a certain temperature [278, 279]. Once the cells have grown and filled with PHB, increasing the temperature from 28 to 42°C causes them to split open and release the polymer. Initially, the yields of PHB from *E. coli* were much lower than those from *A. eutrophus*; yields up to 90% of the cell dry weight have now been reported [280], as well as the preparation of PHBV by a mutant strain.

A very recent development is the transfer of the PHB genes into plants to produce the polymer [189, 276, 281]. The eventual goal would be to modify plants such as corn or potatoes so they produce a thermoplastic biopolymer instead of starch. However, there are many hurdles to be overcome before this can be accomplished; inserting the genes is simple relative to the difficulties of regulating them in the plants. Once the plants can produce PHB, it will be necessary to turn off the production of starch so that the plants can devote all their energy to PHB production. Isolation of the polymer may be more difficult from plant cells, and, in the case of PHB-producing corn, reproduction of the plants becomes a problem, as the plants cannot metabolize PHB to allow seeds to grow.

The simplest approach to obtaining PHB production in plants might be to start with potatoes or turnips. In these plants, the tubers or roots serve only as storage organs, and are not necessary to reproduction. Therefore, it should be possible to modify the potato so that the tuber manufactures polymer in place of starch, but the rest of the potato physiology functions in the usual manner. The first step in this process has already been achieved, with the production of PHB in *Arabidopsis thaliana* [281, 282], a small weed related to the mustard plant. Eventually, PHA could be economically produced by farmers on a very large scale. Transgeneticists suggest that the rapeseed plant, which yields large amounts of oil on a commercial basis, may be the most suitable plant for this purpose [281].

4.9.5   Commercial developments

Active development of PHBV products has been facilitated by the availability of pilot quantities from Zeneca (formerly ICI Biopolymers Division), Billingham, UK. Their stated production capacity is 660 000 lb/ year. The process is similar to the flow chart shown in Figure 4.6. The product is potentially available as a concentrated suspension of granules (latex), a spray-dried powder or formulated pellets. Another reported pilot operation started by Chemie-Linz in Austria uses a solvent extraction process [283, 284] to isolate the polymer; to date, Chemie-Linz has produced several tons of PHB by fermentation with *Alcaligenes latus*, but preparation of PHBV, P(3-HB-co-4HB) and P(3HB-co-4HB-co-3HV) remains at lab scale. Chemie-Linz technology proposes the use of biologically safe nucleating agents and plasticizers to achieve a range of properties for PHB comparable to the PHBV copolyesters [283].

The latex product has been proposed as a replacement for presently used commercial latex polymer applications as a binder or coating in the paper and fibre construct field [117, 192, 194]. Another novel way of depositing PHAs as a supported film is direct electrostatic coating of PHA powder on a moving paper web, followed by fusing and calendaring. Good moisture barrier properties of coated paper have been reported both for latex and powder coating, with retention of biodegradability and recyclability for the paper fibres [192, 193]. Similarly, melt extrusion of formulated pellets for paper coating is under active development to produce biodegradable cups and sandwich wrap for the fast food market [285]. Reports regarding the biodegradability and barrier properties of such films are increasingly available [194].

The most promising use, to date, has been the biodegradable blow-moulded bottle for the shampoo industry. Modern hair shampoo itself is usually biodegradable, hence the biodegradable bottle is a natural environmental complement. Similar packaging applications for oil lubricants or domestic liquid detergents are probably delayed for cost reasons rather than for feasibility. The laminate of paper coated with PHBV may be an attractive approach to lower the cost for packaging applications. Fabrication technology is well-developed in both the paper-coating and packaging fields.

Since, at present, PHBV is the only successful development as a thermoplastic which is both biodegradable and biocompatible, the principal reason for delay in its widespread use must be due to economic factors. Meanwhile, numerous other semi-biodegradable thermoplastics of lower cost are vying for the potential market; these are often blends of gelatinized starch with synthetic thermoplastics. Noteworthy is the thermoplastic poly(ε-caprolactone), the biodegradability and thermoplasticity of which are comparable with PHBV. However, this material is of lower molecular weight, has a lower melting point than PHBV, and is less versatile in terms of overall physical properties.

Acknowledgements

The authors are grateful to Dr Maria Elena Nedea, Dr Christian Lauzier and Dr J. Joseph Jesudason for providing a starting point for this review. Thanks are also due to the Natural Sciences and Engineering Research Council of Canada and Xerox Corp. for financial support.

References

1. Sharp, D. W. A. (1985) *The Penguin Dictionary of Chemistry*, Penguin Books, Harmondsworth, Middlesex, England, p. 320.
2. Dawes, E. A.and Senior, P. J. (1973) *Adv. Microb. Physiol.*, **10**, 203–266.
3. Holmes, P. A. (1988) In *Developments in Crystalline Polymers* (ed. D. C. Basset), Elsevier, New York, Vol. 2, pp. 1–65.
4. Anderson, A. J. and Dawes, E. A. (1990) *Microbiol. Rev.*, **54**(4), 450–472.
5. Doi, Y. (1990) *Microbial Polyesters*, VCH Publishers, New York.
6. Steinbüchel, A. (1991) In *Biomaterials* (ed. D. Byrom), Macmillan Publishers, London, pp. 123–213.
7. Inoue, Y. and Yoshie, N. (1992) *Prog. Polym. Sci.*, **17**, 517–610.
8. Beijerinck, M. W. (1901) *Bakt.*, **11**, 650; cited by Lemoigne, M. (1943) *Compt. Rend.*, **217**, 557.
9. Lemoigne, M. (1925) *C. R. Acad. Sci.*, **180**, 1539.
10. Lemoigne, M. (1925) *Ann. Inst. Pasteur*, **39**, 144.
11. Lemoigne, M. (1926) *Bull. Soc. Chim. Biol.*, **8**, 770.
12. Lemoigne, M. (1927) *Ann. Inst. Pasteur*, **41**, 148.
13. Kepes, A. and Péaud Lenoël, C. (1952) *Bull. Soc. Chim. Biol.*, **34**, 563–575.
14. Weibull, C. (1953) *J. Bacteriol.*, **66**, 696.

15. Williamson, D. H. and Wilkinson, J. F. (1958) *J. Gen. Microbiol.*, **19**, 198.
16. Macrae, R. M. and Wilkinson, J. F. (1958) *J. Gen. Microbiol.*, **19**, 210–222.
17. Merrick, J. M. and Douderoff, M. (1961) *Nature (London)*, **189**, 890–892.
18. Merrick, J. M. (1978) In *Photosynthetic Bacteria* (eds R. K. Clayton and W. R. Sistrom), Plenum Press, New York, pp. 199–219.
19. Dawes, E. A. (1986) *Microbial Energetics*, Blackie, Glasgow, pp. 145–165.
20. Wilkinson, J. F. (1963) *J. Gen. Microbiol.*, **32**, 171–176.
21. Doudoroff, M. and Stanier, R. Y. (1959) *Nature* (London), **183**, 1440–1442.
22. Baptist, J. N. (1962) US Patent Application 3 036 959.
23. Baptist, J. N. (1962) US Patent Application 3 044 942.
24. Baptist, J. N. and Werber, F. X. (1964) *SPE Trans.*, **4**, 245.
25. Baptist, J. N. (1964) US Patent Application 3 121 669.
26. Baptist, J. N. and Werber, F. X. (1965) US Patent Application 3 182 036.
27. Baptist, J. N. and Ziegler, J. B. (1965) US Patent Application 3 225 766.
28. Baptist, J. N. and Werber, F. X. (1963) US Patent Application 3 107 172.
29. Holmes, P. A. (1985) *Phys. Technol.*, **16**, 32–36.
30. Howells, E. R. (1982) *Chem Ind.*, (15), 508–511.
31. Holmes, P. A., Wright, L. F. and Collins, S. H. (1982) European Patent Application 0 052 459.
32. King, P. P. (1982) *J. Chem. Tech. Biotechnol.*, **32**, 2–8.
33. Anon. (1987) *Chemistry in Britain*, **23**(12), 1157.
34. Anon. (1990) *New Scientist*, **126**, 36.
35. Ellar, D., Lundgren, D. G., Okamura, K. and Marchessault, R. H. (1968) *J. Mol. Biol.*, **35**, 489.
36. Ward, A. C., Rowley, B. and Dawes, E. A. (1977) *J. Gen. Microbiol.*, **102**, 61.
37. Müller, H.-M. and Seebach, D. (1993) *Angew. Chem. Internat. Ed. Eng.*, **32**, 477–502.
38. Reusch, R. N. and Sadoff, H. L. (1983) *J. Bacteriol.*, **156**(2), 778–788.
39. Reusch, R. N., Hiske, T. W. and Sadoff, H. L. (1986) *J. Bacteriol.*, **168**(2), 553–562.
40. Reusch, R. N. and Sadoff, H.L. (1988) *Proc. Natl. Acad. Sci. USA*, **85**, 4176–4180.
41. Reusch, R. N. (1989) *Proc. Soc. Exp. Biol. Med.*, **191**, 377–381.
42. Reusch, R. N., Sparrow, A. W. and Gardiner, J. (1992) *Biochim. Biophys. Acta*, **1123**, 33–40.
43. Reusch, R. N. (1992) *FEMS Microbiology Reviews*, **103**, 119–130.
44. Ritchie, G. A. F. and Dawes, E. A. (1969) *Biochem. J,*, **112**, 803–805.
45. Moskowitz, G. J. and Merrick, J. M. (1969) *Biochemistry*, **8**(7), 2748–2755.
46. Ritchie, G. A. F., Senior, P. J. and Dawes, E. A. (1971) *Biochem. J.*, **121**, 309–316.
47. Senior, P. J. and Dawes, E. A. (1971) *Biochem. J.*, **125**, 55.
48. Senior, P. J. and Dawes, E. A. (1973), *Biochem. J.*, **134**, 225–238.
49. Fukui, T., Yoshimoto, A., Matsumoto, M., Hosokawa, S., Saito, T., Nishikawa, H. and Tomita, K. (1976) *Arch. Microbiol.*, **110**, 149–156.
50. Jackson, F. A. and Dawes, E. A. (1976) *J. Gen. Microbiol.*, **97**, 303.
51. Stanier, R. Y., Adelberg, E. A. and Ingraham, J. L. (1976) *The Microbial World, 4th edn* Prentice-Hall, Englewood Cliffs, New Jersey, Chapter 11.
52. Nishimura, T., Saito, T. and Tomita, K. (1978) *Arch. Microbiol.*, **116**, 21–27.
53. Shuto, H., Fukui, T., Saito, T., Shirakura, Y. and Tomita, K. (1981) *Eur. J. Biochem.*, **118**, 53–59.
54. Doi, Y., Kunioka, M., Nakamura, Y. and Soga, K. (1987) *Macromolecules*, **20**, 2988–2991.
55. Haywood, G. W., Anderson, A.J., Chu, L. and Dawes, E.A. (1988) *Biochem. Soc. Trans.*, **16**, 1046–1047.
56. Craig, J. B. and Lloyd, D. R. (1984) *Proceedings of CHEMRAWN III Conference*, The Hague, IUPAC, SECTION 3 IV, Alternative Carbon Sources.
57. Lundgren, D. G., Alper, R., Schnaitman, C. and Marchessault, R. H. (1965) *J. Bacteriol.*, **89**(1), 245–251.
58. Okamura, K. (1967) PhD Thesis, State University of New York, College of Forestry, Syracuse, NY.
59. Marchessault, R. H., Okamura, K. and Su, C. J. (1970) *Macromolecules*, **3**, 735.
60. Kawaguchi, Y. and Doi, Y. (1992) *Macromolecules*, **25**, 2324–2329.

61. Doi, Y., Kawaguchi, Y., Koyama, N., Nakamura, S., Hiramitsu, M., Yoshida, Y. and Kimura, H. (1992) *FEMS Microbiology Reviews*, **103**, 103–108.
62. Wallen, L. L. and Rohwedder, W. K. (1974) *Environ. Sci. Technol.*, **8**, 576–579.
63. Capon, R. J., Dunlop, R. W., Ghisalberti, E. L. and Jefferies, P. R. (1983) *Phytochemistry*, **22**, 1181,
64. Findlay, R. H. and White, P. C. (1983) *Appl. Environ. Microbiol.*, **45**, 71.
65. Odham, G., Tunlid, A., Westerdahl, G. and Maarden, P. (1986) *Appl. Environ. Microbiol.*, **52**(4), 905–910.
66. Bluhm, T. L., Hamer, G. K., Marchessault, R. H., Fyfe, C. A. and Veregin, R. P. (1986) *Macromolecules*, **19**, 2871.
67. Byrom, D. (1987) *Trends Biotechnol.*, **5**, 246–250.
68. Doi, Y., Kunioka, M., Nakamura, Y. and Soga, K. (1986) *J. Chem. Soc. Chem. Commun.*, (23), 1696–1697.
69. Holmes, P. A. and Collins, S. H. (1982) Japan Kokai No. 15 03 93.
70. Bloembergen, S., Holden, D. A., Hamer, G. K., Bluhm, T. L. and Marchessault, R. H. (1986) *Macromolecules*, **19**(11), 2865–2871.
71. Doi, Y., Tamaki, A., Kunioka, M. and Soga, K. (1987) *J. Chem, Soc. Chem. Commun.*, (21), 1635.
72. De Smet, M. J., Eggink, G., Witholt, B., Kingma, J. and Wynbert, H. (1983) *J. Bacteriol.*, **154**, 870.
73. Brandl, H., Gross, R. A., Lenz, R. W. and Fuller, R. C. (1988) *Appl. Environ. Microbiol.*, **54**(8), 1977.
74. Gross, R. A., De Mello, C., Lenz, R. W., Brandl, H. and Fuller, R. C. (1989) *Macromolecules*, **22**, 1106–1115.
75. Barnard, G. N. and Sanders, J. K. M. (1989) *J. Biol. Chem.*, **264**(6), 3286–3291.
76. Davis, J. B. (1964) *Appl. Microbiol.*, **12**, 301–304.
77. Ramsay, B. A., Ramsay, R. A. and Cooper, D. G. (1989) *Appl. Environ. Microbiol.*, **55**, 584.
78. Steinbüchel, A., Debzi, E.-M., Marchessault, R. H. and Timm, A. (1993) *Appl. Microbiol. Biotechnol.*, **39**, 443–449.
79. Haywood, G. W., Anderson, A. J., Williams, D. R. and Dawes, E. A. (1991) *Internat. J. Biol. Macromol.*, **13**, 83.
80. Stanier, R. Y., Palleroni, N. J. and Doudoroff (1966) *J. Gen. Microbiol.*, **43**, 159.
81. Alper, R., Lundgren, D. G., Marchessault, R. H. and Cote, W. A. (1963) *Biopolymers*, **1**, 545–556.
82. Walker, J., Whitton, J. R. and Alderson, B. (1982) European Patent Application EP 0 046 017.
83. Solvay and Cie (1979) European Patent Application EP 14 490.
84. Barham, P. J. and Selwood, A. (1982) European Patent Application EP 0 058 480.
85. Barham, P. J. and Selwood, A. (1982) US Patent Application 4 391 766.
86. Lafferty, R. M. and Heinzle, E. (1977) *Chem. Rundshau*, **30**(41), 15–16.
87. Nuti, M. P., De Bertoldi, M. and Lepidi, A. A. (1972) *Can J. Microbiol.*, **18**, 1257.
88. Berger, E., Ramsay, B. A., Ramsay, J. A., Chavarie, C. and Braunegg, G. (1989) *Biotechnol. Tech.*, **3**(4), 227.
89. Ramsay, J. A., Berger, E., Ramsay, B. A. and Chavarie, C. (1989) *Biotechnol. Tech.*, **4**(4), 221.
90. Merrick, J. M. and Doudoroff, M. (1964) *J. Bacteriol.*, **88**(1), 60–71.
91. Griebel, R., Smith, Z. and Merrick, J. M. (1968) *Biochemistry*, **7**(10), 3676–3681.
92. Kawaguchi, Y. and Doi, Y. (1990) *FEMS Microbiol. Lett.*, **70**, 151.
93. Holmes, P. A. and Lim, G. B. (1985) European Patent Application EP 145 233.
94. Holmes, P. A. and Jones, E. (1980) European Patent Application EP 46 335.
95. Barham, P. J. and Keller, A. (1986) *J. Polym. Sci., Polym. Phys. Ed.*, **24**, 69.
96. Basta, N. (1984) *High Technology*, February, 67–71.
97. Geil, P. H. (1963) *Polymer Single Crystals (Polymer Reviews, No. 5)*, John Wiley, New York.
98. Marchessault, R. H., Morikawa, H., Revol, J. F. and Bluhm, T. L. (1984) *Macromolecules*, **17**, 1882.

99. Yokouchi, M., Chatani, Y., Tadokoro, H., Teranishi, K. and Tani, H. (1973) *Polymer*, **14**, 267–272.
100. Okamura, K. and Marchessault, R. H. (1967) In *Conformation of Biopolymers* (ed. G. M. Ramachandran), Academic Press, New York, Vol. 2, p. 709.
101. Cornibert, J. and Marchessault, R. H. (1972) *J. Mol. Biol.*, **71**, 735–756.
102. Marchessault, R. H., Coulombe, S., Morikawa, H., Okamura, K. and Revol, J. F. (1981) *Can. J. Chem.*, **59**, 38.
103. Allegra, G. and Bassi, I. W. (1969) *Adv. Polym. Sci.*, **6**, 549.
104. Adsetts, J. (1975) In *Polymer Handbook, 2nd Edn* (eds J. Brandrup and E. H. Immergut), John Wiley, New York, p. V–23.
105. Bloembergen, S., Holden, D. A., Bluhm, T. L., Hamer, G. K. and Marchessault, R. H. (1987) *Macromolecules*, **20**, 3086–3089; (1989) *Macromolecules*, **22**, 1663.
106. Mitomo, H., Barham, P. J. and Keller, A. (1988) *Polym. Commun.*, **29**, 112.
107. Kamiya, N., Yamamoto, Y., Inoue, Y., Chûjô, R. and Doi, Y. (1989) *Macromolecules*, **22**, 1676.
108. Kunioka, M., Tamaki, A. and Doi, Y. (1989) *Macromolecules*, **22**, 694.
109. Revol, J. F., Chanzy, H. D., Deslandes, Y. and Marchessault, R. H. (1989) *Polymer*, **30**, 1973.
110. Scandola, M., Ceccorulli, G. and Doi, Y. (1990) *Internat. J. Biol. Macromol.*, **12**, 112.
111. Orts, W. J., Marchessault, R. H. and Bluhm, T. L. (1991) *Macromolecules*, **24**, 6435.
112. Scandola, M., Ceccorulli, G., Pizzoli, M. and Gazzano, M. (1992) *Macromolecules*, **25**, 1405–1410.
113. Yoshie, N., Sakurai, M., Inoue, Y. and Chûjô, R. (1992) *Macromolecules*, **25**, 2046–2048.
114. Barham, P. J., Keller, A., Otun, E. L. and Holmes, P. A. (1984) *J. Mater. Sci.*, **19**, 2781–2794.
115. Organ, S. J. and Barham, P. J. (1991) *J. Mater. Sci.*, **28**, 1368–1374.
116. Bonthrone, K. M., Clauss, J., Horowitz. D. M., Hunter, B. K. and Sanders, J. K. M. (1992) *FEMS Microbiology Reviews*, **103**, 269–278.
117. Marchessault, R. H., Monasterios, C. J. and Lepoutre, P. (1990) In *Novel Biodegradable Microbial Polymers* (ed. E. A. Dawes), Kluwer, Dordrecht pp. 97–112.
118. Boatman, E. S. (1964) *J. Cell. Biol.*, **20**, 297.
119. Lundgren, D. G., Pfister, R. M. and Merrick, J. M. (1964), *J. Gen. Microbiol.*, **34**, 441.
120. Pfister, R. M. and Lundgren, D. G. (1964) *J. Bacteriol.*, **88**(4), 1119.
121. Wang, W. S. and Lundgren, D. G. (1969) *J. Bacteriol.*, **97**(2), 947.
122. Nickerson, K. W. (1982) *Appl. Environ. Microbiol.*, **43**(5), 1208.
123. Fuller, R. C., O'Donnell, J. P., Saulnier, J., Redlinger, T. E., Foster, J. and Lenz, R. W. (1992) *FEMS Microbiol. Rev.*, **103**, 279–288.
124. Merrick, J. M. (1965) *J. Bacteriol.*, **90**(4), 965.
125. Merrick, J. M., Lundgren, D. G. and Pfister, R. M. (1965) *J. Bacteriol.*, **89**(1), 234.
126. Hippe, H. and Schlegel, H. G. (1967) *Arch. Mikrobiol.*, **56**, 278–299.
127. Griebel, R. J. and Merrick, J. M. (1971) *J. Bacteriol.*, **108**(2), 782–789.
128. Mayer, F. (1992) *FEMS Microbiology Reviews*, **103**, 265–268.
129. Smith, P. and Lemstra, P. J. (1980) *J. Mater. Sci.*, **15**, 505.
130. Smith, P. and Lemstra, P. J. (1980) *Colloid Polym. Sci.*, **258**, 891.
131. Smith, P. and Lemstra, P. J. (1980) *Polymer*, **21**, 1341.
132. Ito, H., Marchessault, R. H. and Manley, R. St. J. (1991) *Polym. Commun.*, **32**(6), 164–167.
133. Lauzier, C. (1992) *Morphology and Crystallization Behaviour of Nascent Poly(3-hydroxybutyrate) Granules*, PhD Thesis, McGill University, Montreal, Quebec, Canada.
134. Dunlop, W. F. and Robards, A. W. (1973) *J. Bacteriol.*, **114**(3), 1271–1280.
135. Lauzier, C., Marchessault, R. H., Smith P. and Chanzy, H. (1992) *Polymer*, **33**(4), 823–827.
136. Lauzier, C., Revol, J.-F., and Marchessault, R. H. (1992) *FEMS Microbiology Reviews*, **103**, 299–310.
137. Horowitz, D. H., Clauss, J., Hunter, B. K. and Sanders, J. K. M. (1993) *Nature*, **363**, 23.
138. De Koning, G. J. M. and Lemstra, P. J. (1992) *Polymer*, **33**(15), 3292–3294.
139. Scherer, G. W. (1993) *J. Non-Crystall. Solids*, **155**, 1–25.
140. Merrick, J. M., Delafield, F. P. and Doudoroff, M. (1962) *Fed. Proc.*, **21**, 228.

141. Hippe, H. (1967) *Arch. Mikrobiol.*, **56**, 248.
142. Nakada, T., Fukui, T., Saito, T., Miki, K., Oji, C., Matsuda, S., Ushijima, A. and Tomita, K. (1981) *J. Biochem.*, **89**, 625–635.
143. Fukui, T., Ito, M. and Tomita, K. (1982) *Eur. J. Biochem.*, **127**, 423–428.
144. Oeding, V. and Schlegel, H. G. (1973) *Biochem, J.*, **134**, 239–248.
145. Schindler, J. and Schlegel, H. G. (1963) *Biochemische Zeitschrift*, **339**, 154–161.
146. Gavard, R., Dahinger, A., Hauttecoeur, B. and Reynaud, C. (1967) *C. R. Acad. Sci.*, **265**, 1557.
147. Tanaka, Y., Saito, T., Fukui, T., Tanio, T. and Tomita, K. (1981) *Eur. J. Biochem.*, **118**, 177–182.
148. Merrick, J. M. and Yu, C. I. (1966) *Biochemistry*, **5**(11), 3563–3568.
149. Delafield, F. P., Cooksey, K. E. and Doudoroff, M. (1965) *J. Biol. Chem.*, **240**(10), 4023.
150. Doi, Y., Segawa, A., Kawaguchi, Y. and Kunioka, M. (1990) *FEMS Microbiol. Lett.*, **67**, 165.
151. Delafield, F. P., Doudoroff, M., Palleroni, N. J., Lusty, C. J. and Contopoulos, R. (1965) *J. Bacteriol.*, **90**(5), 1455.
152. Nakayama, K., Saito, T., Fukui, T., Shirakura, Y. and Tomita, K. (1985) *Biochim. Biophys. Acta*, **827**, 63.
153. Tanio, T., Fukui, T., Shirakura, Y., Saito, T., Tomita, K., Kaiho, T. and Masamune, S. (1982) *Eur. J. Biochem.*, **124**, 71–77.
154. Shirakura, Y., Fukui, T., Saito, T., Okamoto, Y., Narikawa, T., Koide, K., Tomita, K., Takemasa, T. and Masamune, S. (1986) *Biochim. Biophys. Acta*, **880**, 46–53.
155. Fukui, T., Narikawa, T., Miwa, K., Shirakura, Y., Saito, T. and Tomita, K. (1988) *Biochim. Biophys. Acta*, **952**, 164,
156. Saito, T., Suzuki, K., Yamamoto, J., Fukui, T., Miwa, K., Tomita, K., Nakanishi, S., Odani, S., Suzuki, J. I. and Ishikawa, K. (1989) *J. Bacteriol.*, **171**(1), 184.
157. Jendrossek, D., Knoke, I., Habibian, R. B., Steinbüchel, A. and Schlegel, H. G. (1993) *J. Environ. Polym. Degrad.*, **1**(1) 53–63.
158. Mukai, K., Yamada, K., and Doi, Y. (1993) *Polym. Degrad. Stab.*, **41**, 85–91.
159. Saito, T., Iwata, A. and Watanabe, T. (1993) *J. Environ. Polym. Degrad.*, **1**(2), 99–105.
160. Henrissat, B., Vigny, B., Buleon, A. and Perez, S. (1988) *Fed. Eur. Biochem. Soc.*, **231**(1), 177–182.
161. Gilkes, N. R., Jervis, E., Henrissat, B., Tekant, B., Miller, R. C. Jr., Warren, R. A. J. and Kilburn, D. G. (1992) *J. Biol. Chem.*, **267**(10), 6743–6749.
162. Kumagai, Y. and Doi, Y. (1992) *Polym. Degrad. Stab.*, **35**, 87.
163. Kumagai, Y., Kanesawa, Y. and Doi, Y. (1992) *Makromol. Chem.*, **193**, 53.
164. Doi, Y., Kanesawa, Y., Kunioka, M. and Saito, T. (1990) *Macromolecules*, **23**, 26.
165. Morikawa, H. and Marchessault, R. H. (1981) *Can. J. Chem.*, **59**, 2306–2313.
166. Hauttecoeur, B., Jolivet, M. and Gavard, R. (1972) *C. R. Acad. Sci., Ser. D*, **274**(19), 2729–2732.
167. Grassie, N., Murray, E. J. and Holmes, P. A. (1984) *Polym Degrad. Stab.*, **6**, 47.
168. Grassie, N., Murray, E. J. and Holmes, P. A. (1984) *Polym. Degrad. Stab*, **6**, 95.
169. Gould, G. S. (1959) In *Mechanism and Structure in Organic Chemistry*, Holt, Rinehart and Wilson, New York, p. 502.
170. Grassie, N., Murray, E. J. and Holmes, P. A. (1984) *Polym. Degrad. Stab.*, **6**, 127.
171. Holland, S. J., Jolly, A. M., Yasin, M. and Tighe, B. J. (1987) *Biomaterials*, **8**(4), 289–295.
172. Miller, N. D. and Williams, D. F. (1987) *Biomaterials*, **8**(2), 129–137.
173. Holland, S. J., Yasin, M. and Tighe, B. J. (1990) *Biomaterials*, **11**(3), 206–215.
174. Yasin, M., Holland, S. J. and Tighe, B. J. (1990) *Biomaterials*, **11**(7), 451–454.
175. Doi, Y., Kanesawa, Y., Kawaguchi, Y. and Kunioka, M. (1989) *Makromol. Chem., Rapid Commun.*, **10**, 227–230.
176. Karlsson, S., Sares, C. and Albertsson, A.-C. (1992) *FEMS Microbiology Reviews*, **103**, 455–456.
177. Kanesawa, Y. and Doi, Y. (1990) *Makromol. Chem., Rapid Commun.*, **11**, 679–682.
178. Matavulj, M., Moss, S. T. and Molitoris, H. P. (1992) *FEMS Microbiology Reviews*, **103**, 465–466.
179. Doi, Y., Kanesawa, Y., Tanahashi, N. and Kumagai, Y. (1992) *Polym. Degrad. Stab.*, **36**, 173–177.

180. ICI Americas Inc. (1988) *PHBV Biodegradable Polyesters: Natural, Thermoplastic, Biodegradable,* ICI promotional material.
181. Reiman, T. (1986) *Can. Res.,* **19**(9), 22–23.
182. Bluhm, T. and Marchessault, R. H. (1988) *Can. Chem. News,* **40**(9), 25–26.
183. Smith, D. K., Belson, N. and Kilp, T. (1990) *A Study of Degradable Plastics,* Queen's Printer for Ontario, Canada.
184. Püchner, P. and Müller, W.-R. (1992) *FEMS Microbiology Reviews,* **103**, 469–470.
185. Swift, G. (1992) *FEMS Microbiology Reviews,* **103**, 339–346.
186. Augusta, J., Müller, R.-J. and Widdecke, H. (1992) *FEMS Microbiology Reviews,* **103**, 477–479.
187. Moore, J. W. (1992) *Modern Plastics,* mid-Dececmber, 58–63.
188. Fukuda, K. (1993) Presented at Gordon Conference on Biodegradable Polymers.
189. Pool, R. (1989) *Science,* **245**, 1187–1189.
190. Ramsay, J. A. and Ramsay, B. A. (1990) *Applied Phycology Forum,* **7**(3), 1–5.
191. McCarthy-Bates, L. (1993) *Plastics World,* March, 22–27.
192. Marchessault, R. H., Lepoutre, P. and Wrist, P. (1991) PCT International Application WO 91 13 207.
193. Marchessault, R. H., Rioux, P. and Saracovan, I. (1993) *Nordic Pulp and Paper Research Journal,* (1), 211–216.
194. Lauzier, C. A., Monasterios, C. J., Saracovan, I., Marchessault, R. H. and Ramsay, B. A. (1993) *Tappi,* **76**(5), 71–77.
195. Fuller, T. J., Marchessault, R. H. and Bluhm, T. L. (1991) US Patent Application 5 004 664.
196. Holland, S. J., Tighe, B. J. and Gould, P. L. (1986) *J. Controlled Release,* **4**(3), 155–180.
197. Kwan, L. and Steber, W. (1991) European Patent Application EP 406 015.
198. ICI (~ 1982) *Biopol, the Unique Biodegradable Thermoplastic from ICI,* promotional material.
199. Pawan, G. L. S. and Semple, S. J. G. (1983) *Lancet,* 1 January, 15–17.
200. Stockmann, H. H. and Ray-Chaudhuri, D. K. (1971) US Patent Application 3 579 549.
201. Trau, M. and Truss, R. W. (1988) European Patent Application EP 293 172.
202. Kubota, M., Nakano, M. and Juni, K. (1988) *Chem. Pharm. Bull.,* **36**(1), 333–337.
203. Korsatko, W., Wabnegg, B., Braunegg, G., Lafferty, R. M. and Strempfl, F. (1983) *Pharm. Ind.,* **45**(5), 525–527.
204. Korsatko, W., Wabnegg, B., Tillian, H. M., Braunegg, G. and Lafferty, R. M. (1983) *Pharm. Ind.,* **45**(10), 1004–1007.
205. Korsatko, W., Wabnegg, B., Tillian, H. M., Egger, G., Pfragner, R. and Walser, V. (1984) *Pharm. Ind,,* **46**(9), 952–954.
206. Trathnigg, B., Wiedmann, V., Lafferty, R. M., Korsatko, B. and Korsatko, W. (1988), *Angew. Makromol. Chem.,* **161**, 1–8.
207. Holmes, P. A. (1985) British UK Patent Application GB 2 160 208.
208. Kamata, T., Numazawa, R. and Kamo, J. (1983) *Jpn. Kokai Tokyo Koho* JP 60 137 402.
209. Heimerl, A., Pietsch, H., Rademacher, K. H., Schwengler, H., Winkeltau, G. and Treutner, K. H. (1989) European Patent Application EP 336 148.
210. Bowald, S. F. and Johansson, E. G. (1990) European Patent Application EP 349 505.
211. Talja, M., Tormala, P., Rokkanen, P., Vainionpaa, S. and Pohjonen, T. (1990) PCT International Application WO 90 04 982.
212. Ando, Y. and Fukada, E. (1984) *J. Polym. Sci., Polym. Phys. Ed.,* **22**, 1821–1834.
213. Fukada, E. (1983) *Quart. Rev. Biophys.,* **16**, 59–87.
214. Pirkle, W. H., Finn, J. M., Schreiner, J. L. and Hamper, B. C. (1981) *J. Am. Chem. Soc.,* **103**, 3964–3966.
215. Seebach, D., Roggo, S. and Zimmermann, J. (1987) In *Stereochemistry of Organic and Bioorganic Transformations Workshop Conferences Hoechst* (eds W. Bartmann and K. B. Sharpless), VCH Verlagsgesellschaft mbH., Vol. 17, pp. 87–126.
216. Seuring, B. and Seebach, D. (1978) *Liebigs Ann. Chem.,* 2044–2073.
217. Sutter, M. A. and Seebach, D. (1983) *Liebigs Ann. Chem.,* 939–949.
218. Seidel, W. and Seebach, D. (1982) *Tet. Lett.,* **23**(2), 159–162.
219. Mori, K. (1977) *Tetrahedron,* **33**, 289–294.

220. Mori, K. and Watanabe, H. (1984) *Tetrahedron*, **40**(2), 299–303.
221. Kloss, M. (1988) *Ger. Offen*. DE 3 711 598.
222. Biedermann, J., Staniszewski, M., Wais, S. and Süßmuth, R. (1992) *FEMS Microbiology Reviews*, **103**, 473–474.
223. Wurmthaler, J. and Müller, W.-R. (1992) *FEMS Microbiology Reviews*, **103**, 475–476.
224. CW Price Report (1993) *Chemical Week*, **152**(23), 72.
225. Suzuki, T., Deguchi, H., Yamane, T., Shimizu, S. and Gekko, K. (1988) *Appl. Microbiol. Biotechnol.*, **27**, 487–491.
226. Page, W. J. (1989) *Appl. Microbiol. Biotechnol.*, **31**, 329–333.
227. Page, W. J. (1992) *FEMS Microbiology Reviews*, **103**, 149–158.
228. Bertrand, J.-L., Ramsay, B. A., Ramsay, J. A. and Chavarie, C. (1990) *Appl. Environ. Microbiol.*, **56**, 3133–3138.
229. Ramsay, J. A., private communication (Ecole Polytechnique, Montreal, Quebec, Canada).
230. Lageveen, R. G., Huisman, G. W., Preusting, H., Ketelaar, P., Eggink, G. and Witholt, B. (1988) *Appl. Environ. Microbiol.*, **54**(12), 2924–2932.
231. Witholt, B., Lageveen, R. G., Huisman, G. W., Preusting, H., Nijenhuis, A., Kingma, J., Tijsterman, A. and Eggink, G. (1988) *Polym. Prepr.*, **29**(1), 592–593.
232. Eggink, G., van der Wal, H. and Huyberts, G. (1990) In *Novel Biodegradable Polymers* (ed. E. A. Dawes), Kluwer, Dordrecht, pp. 441–444.
233. Fritzsche, K., Lenz, R. W. and Fuller, R. C. (1990) *Internat. J. Biol. Macromol.*, **12**, 85–91.
234. Fritzsche, K., Lenz, R. W. and Fuller, R. C. (1990) *Internat. J. Biol. Macromol.*, **12**, 92–102.
235. Lenz, R. W., Kim, B. -W., Ulmer, H. W. and Fritzsche, K. (1990) In *Novel Biodegradable Polymers* (ed. E. A. Dawes), Kluwer, Dordrecht, pp. 23–35.
236. Preusting, H., Nijenhuis, A. and Witholt, B. (1990) *Macromolecules*, **23**, 4220–4224.
237. Morin, F. G. and Marchessault, R. H. (1992) *Macromolecules*, **25**, 576.
238. Marchessault, R. H., Monasterios, C. J., Morin, F. G. and Sundarajan, P. R. (1990) *Internat. J. Biol. Macromol.*, **12**, 158–165.
239. Kim, Y. B., Lenz, R. W. and Fuller, R. C. (1992) *Macromolecules*, **25**, 1852–1857.
240. Lenz, R. W., Kim, Y. B. and Fuller, R. C. (1992) *FEMS Microbiology Reviews*, **103**, 207–214.
241. Doi, Y., Kunioka, M., Nakamura, Y. and Soga, K. (1988) *Macromolecules*, **21**, 2722.
242. Kunioka, M., Kawaguchi, Y. and Doi, Y. (1989) *Appl. Microbiol. Biotechnol.*, **30**, 569.
243. Doi, Y., Segawa, A. and Kunioka, M. (1990) *Internat. J. Biol. Macromol.*, **12**, 106.
244. Nakamura, S., Doi, Y. and Scandola, M. (1992) *Macromolecules*, **25**, 4237.
245. Kunioka, M., Nakamura, Y. and Doi, Y. (1988) *Polym. Commun.*, **29**, 74.
246. Nakamura, S., Kunioka, M. and Doi, Y. (1991) *Macromolecular Reports*, **A28**(Suppl. 1), p. 15.
247. Johns, D. B., Lenz, R. W. and Luecke, A. (1984) In *Ring-Opening Polymerization* (eds. K. J. Ivin and T. Saegusa), Elsevier, New York, Vol. 1, pp. 461–521.
248. Knobloch, F. W., Hockessin and Statton, W. O. (1967) US Patent Application 3 299 171.
249. Gross, R. A., Zhang, Y., Konrad, G. and Lenz, R. W. (1988) *Macromolecules*, **21**, 2657.
250. Benvenuti, M. and Lenz, R. W. (1991) *J. Polym. Sci., Polym. Chem.*, **29**, 793.
251. Iida, M., Araki, T., Teranishi, K. and Tani, H. (1977) *Macromolecules*, **10**, 275.
252. Zhang, Y., Gross, R. A. and Lenz, R. W. (1990) *Macromolecules*, **23**, 3206.
253. Bloembergen, S., Holden, D. A., Bluhm, T. L., Hamer, G. K. and Marchessault, R. H. (1989) *Macromolecules*, **22**, 1656.
254. Hocking, P. J. and Marchessault, R. H. (1993) *Polym. Bull.*, **30**, 163–170.
255. Kemnitzer, J. E., McCarthy, S. P. and Gross, R. A. (1993) *Macromolecules*, **26**, 1221–1229.
256. Kemnitzer, J. E., McCarthy, S. P. and Gross, R. A. (1992) *Macromolecules*, **25**, 5927–5934.
257. Jesudason, J. J., Marchessault, R. H. and Saito, T. (1993) *J. Environ. Polym. Degrad.*, **1**(2), 89–98.
258. Avella, M. and Martuscelli, E. (1988) *Polymer*, **29**, 1731–1737.

259. Greco, P. and Martuscelli, E. (1989) *Polymer*, **30**, 1475–1483.
260. Marand, H. and Collins, M. (1990) *Polym. Prepr.*, **31**, 552.
261. Dave, B., Parikh, M., Reeves, M. S., Gross, R. A. and McCarthy, S. P. (1990) *Polym. Mater. Sci. Eng.*, **63**, 726.
262. McCarthy, S. P. and Gross, R. (1991) *Proceedings of Environmentally Degradable Polymers: Technical, Business, and Public Perspectives*, Chelmsford, MA.
263. Yasin, M., Holland, S. J., Jolly, A. M. and Tighe, B. J. (1989) *Biomaterials*, **10**(6), 400–412.
264. Langlade, V. (1991) *Étude de la préparation et de la biodégradation de mélanges de poly-β-hydroxyalcanoates avec des polymères conventionnels*, MSc Thesis, Université de Montréal, Montreal, Quebec, Canada.
265. Koenig, M. F. and Huang, S. J. (1992) *ACS Polym. Mater. Sci. Eng.*, **67**, 290–291.
266. Marchessault, R. H., Bluhm, T. L., Deslandes, Y., Hamer, G. K., Orts, W. J., Sundarajan, P. R., Taylor, M. G., Bloembergen, S. and Holden, D. A. (1988) *Makromol. Chem., Macromol. Symp.*, **19**, 235.
267. Kumagai, Y. and Doi, Y. (1992) *Makromol. Chem., Rapid Commun.*, **13**, 179.
268. Pearce, R., Jesudason, J., Orts, W., Marchessault, R. H. and Bloembergen, S. (1992) *Polymer.*, **33**(21), 4647–4649.
269. Scandola, M., Ceccorulli, G. and Pizzoli, M. (1992) *Macromolecules*, **25**, 6441–6446.
270. Buchanan, C. M., Gedon, S. C., White, A. W. and Wood, M. D. (1992) *Macromolecules*, **25**, 7373–7381.
271. Lotti, N. and Scandola, M. (1992) *Polym. Bull.*, **29**, 407–413.
272. Slater, S. C., Voige, W. H. and Dennis, D. E. (1988) *J. Bacteriol.*, **170**(10), 4431–4436.
273. Schubert, P., Steinbüchel, A. and Schlegel, H. G. (1988) *J. Bacteriol.*, **170**(12), 5837–5847.
274. Peoples, O. P. and Sinskey, A. J. (1989) *J. Biol. Chem.*, **264**(26), 15293–15297.
275. Peoples, O. P. and Sinskey, A. J. (1989) *J. Biol. Chem.*, **264**(26), 15298–15303.
276. McWilliams, G. (1991) *Business Week*, August 19, 110–111.
277. Schubert, P., Krüger, N. and Steinbüchel, A. (1991) *J. Bacteriol.*, **173**(1), 168–175.
278. Witte. A. and Lubitz, W. (1989) *Eur. J. Biochem.*, **180**, 393–398.
279. Lubitz, W. (1991) European Patent Application EP 435 028.
280. Fidler, S. and Dennis, D. (1992) *FEMS Microbiology Reviews*, **103**, 231–236.
281. Poirier, Y., Dennis, D., Klomparens, K., Nawrath, C. and Somerville, C. (1992) *FEMS Microbiology Reviews*, **103**, 237–246.
282. Poirier, Y., Dennis, D. E., Klomparens, K. and Somerville, C. (1992) *Science*, **256**, 520–523.
283. Hänggi, U. J. (1990) In *Novel Biodegradable Microbial Polyesters* (ed. E. A. Dawes), Kluwer, Dordrecht, pp. 65–70.
284. Hrabak, O. (1992) *FEMS Microbiology Reviews*, **103**, 251–256.
285. Doi, Y. private communication (RIKEN, Tokyo, Japan).
286. Jesudason, J. J. (1993) *Synthetic Analogues of Bacterial Polyesters: Preparation and Properties*, PhD Thesis, McGill University, Montreal, Quebec, Canada.
287. Jesudason, J. J. and Marchessault, R. H. (1993) *Macromolecules* (in press).

5 Recycling technology for biodegradable plastics

C. R. ARMISTEAD

5.1 Introduction

This chapter describes work done by the Manchester Packaging Co., USA regarding recycling in-plant degradable polyethylene/starch (PE/S) blown film. This work was due to economic necessity not academic curiosity. We believe that degradable plastics will play a positive role in solid waste management. The primary obstacles to acceptance are traditional. First, general ignorance of these new materials and their performance limitations creates public disappointment when performance expectations are not achieved. Secondly, economics is a problem – new technology is usually initially more expensive than existing technology, as is the case with PE/S products and other degradable polymers under development. In our experience, the public expects degradable materials to last as long as needed, then to disappear harmlessly as soon as the material knows it has been discarded. This is a demanding expectation to say the least. Our philosophy is to recommend degradable material whenever it is probable and logical to expect the product made from the PE/S to be disposed in an environment that supports degradation and no other alternative means of handling the disposed material exists; or whenever degradability is a functional requirement of intended use. Of course, it does not make sense to put degradables in landfills. Current landfill technology does not support degradation of any material to any appreciable degree in the time expectation of the public [1]. This is a landfill design problem, however, not a failure of degradable materials to perform. It is amusing to hear and read that plastics will last up to 500 years in landfills. Modern olefin plastics are barely 50 years old and most landfills are younger. 'Experts' are making 500 year forecasts on plastics' degradability and yet tomorrow's weather cannot be predicted accurately. Both the weather and plastic degradation are complex and incompletely understood processes. Why are people so willing to accept the predictions regarding one but not the other?

5.2 Conventional recycling

5.2.1 *Economic incentive*

Conventional polyethylene blown film is manufactured on a single screw machine designed to produce as uniform a polymer melt as possible at economical production rates. Almost from the beginning, processors have recycled in-plant waste into saleable product. The primary reason was and is economic. Using a simple rotary blade grinding machine, scrap polyethylene film can be ground into fluff or densified into granular form. With the additional step of repelletizing, a denser and easier handling pellet similar to virgin resin can be achieved. This fluff, densified granular, or repelletized material is referred to as 'repro'. This repro material is usually blended back into virgin polyethylene and blown into film again. For some uses, such as trash bags, blown film is made entirely from repro. The cost of reprocessing film scrap is generally 30–50% cheaper than virgin resin costs, depending on cyclical market conditions. For every kilogram of reprocessed material used, a kilogram of higher priced virgin resin is not needed. Additionally, the processor does not have to pay to dispose of the waste film in a landfill even at cheap rates such as $22.00 (US) per metric ton. This economic incentive to polyethylene processors has made recycling in-plant scrap common place. It has even resulted in the formation of a secondary market in reprocessed resin.

5.2.2 *Recycling problems*

While recycling in-plant waste offers economic advantages, it also presents challenges to the processor. Polyethylene is produced in many different grades with varying properties tailored toward the end use of the film. Some grades are more expensive than others and some grades will not readily mix with other grades to produce acceptable quality blown film. Additionally, all polyethylene grades will gradually deteriorate in quality the more times they are heated and reprocessed. The polymer molecules suffer crosslinking and oxidation. Loss of tensile strength, impact resistance and gloss or clarity occur on repeated reprocessing. Since most polyethylene blown film is used in packaging, much of it is imprinted with multicolored advertising, instructions and graphics. When this printed film is reprocessed, the residual inks color the film and add to the loss in properties. Clear printed film upon reprocessing becomes black or brown or some other color depending on the amount and colors of ink with which it was originally printed. To make matters even worse, many films are both pigmented and printed. Pigment and its carrier resin can also reduce physical properties of the resulting film. Most carrier resins used in color concentrates are 20 melt index (MI) to enhance compounding of the pigment and to aid in dispersal of color with film grade resins of 0.25 to 2.0 MI in the blown film extruder.[1] The only advantage the processor has is control of the scrap. In-plant scrap can be traced, identified, segregated and selectively reprocessed because the processor knows what is being produced. This is one reason post-consumer recycling is more difficult and expensive than proponents first thought. With post-consumer scrap, it is more difficult to know of what resin the film is made. Post-consumer scrap is likely to be a mixture of film resins which may not be compatible. Separating these incompatible scraps can be time consuming and expensive, if even possible. If the different grades are not separated, the resulting product is either inferior in properties or useless.

5.3 Degradables complicate recycling

5.3.1 Polyethylene/corn starch film

To understand the problems in recycling polyethylene/corn starch material, it is necessary to understand the properties and problems of the material and its manufacture. When Griffin introduced the concept of degradable films by incorporating corn starch with a carrier resin, the idea of recycling such a product was not foremost in anyone's mind [2]. The product was intended to be biodegradable in the environment. Corn starch manufacturers were interested in a new market for corn starch and the large volume polyethylene packaging industry offered potential. In 1989, *Modern Plastics* magazine estimated the US market at 2 billion pounds (1 million metric tons) for grocery, trash and merchandise bags made of plastics [3]. If only a modest portion of it used 6% corn starch as was originally expected, millions of bushels of corn would be consumed. Of course, this corn starch would displace millions of tons of virgin resin about which the resin companies were not enthusiastic. The potential that many more pounds of resin might be consumed in new applications for degradables using corn starch, more than offsetting the 6% lost to starch, was not considered by the resin manufacturers. This issue became a point of contention for the resin manufacturers and was one reason they did not support degradable technology.

5.3.1.1 Processing with conventional equipment. To manufacture polyethylene with corn starch, masterbatch containing about 45 wt% modified corn starch in a 20 MI polyethylene carrier resin was blended with conventional polyethylene. The blended material was fed to an unmodified blown film extruder in the same manner as conventional resin mixes. The original formulation supplied by Archer-Daniels-Midland Company or St. Lawrence Starch, required use of 14 wt% corn starch masterbatch to 86 wt%

polyethylene.[2] This resulted in a finished blown film with a 6 wt% starch content which was the optimum level recommended by research done by the masterbatch producers. Later, Fully Compounded Plastics Inc. of Decatur, Illinois, introduced a masterbatch containing 50 wt% corn starch in an ethylene methyl acrylate polyethylene carrier resin.[3] FCP material required only 12 wt% masterbatch to 88 wt% polyethylene to achieve 6 wt% starch in the finished film. We used conventional 2½ inch (63 mm) and 3½ inch (89 mm) single screw extruders with a 24:1 length to diameter ratio. The machines were Gloucester Engineering Company (GEC) and Sterling Extruder design.[4] Screw design was Sterlex® double flite barrier type, designed to process mixtures of high pressure, low density polyethylene (LDPE) and linear low density polyethylene (LLDPE). The blown film dies were GEC, Sterling or Sano Inc.[5] low pressure non-adjustable design with 0.032 to 0.080 inch (0.8–2.0 mm) die gap; die sizes ranged from 3 to 14 inch (76–356 mm) diameter. Air rings used were various designs of dual-lipped configuration with 45 to 50°F (7–10°C) chilled air used for cooling. The blown film process was found to be fairly tolerant to the addition of corn starch masterbatch, but there were problems.

5.3.1.2 Starch particle effects. First, with initial formulations, starch particle size affected melt strength. Too large starch particles acted as stress sites for blow hole formation in the melt region of the bubble below the frost line, causing loss of bubble size or complete collapse of the bubble. Agglomeration of starch particles was seen which appeared as large specks in the film which had the same effect as too large starch particles. This was a source of scrap and down time which is not economical in the blown film business. This problem was overcome by proper selection, preparation and compounding of the starch by the masterbatch producers. It was also found that starch added to polyethylene affected film yield and thickness measurement. Corn starch is denser than polyethylene, combining the two produced material with an average density higher than normal polyethylene. Hence a roll of PE/S film of a given length would weigh more than a roll of conventional poly of the same length. The difference was ~3% and had to be compensated for to make sure enough film footage was produced to yield correct bag quantities when converted. Additionally, the starch particles in the film produced a surface texture different from conventional film. Some starch particles protrude above the general polyethylene matrix (Figure 5.1). Using a micrometer to measure film thickness resulted in readings thicker than the true average. This was a problem in that too little polyethylene in the mixture would yield a film weaker than desired since true thickness was masked by the starch particles protruding. Griffin's research detailed how to correct this by back calculating film thickness from the weight of a known area of film and the film resin mix

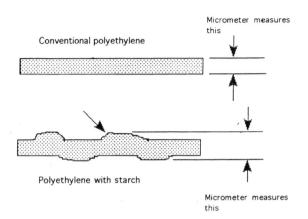

Figure 5.1 Polyethylene film containing starch masterbatch measures thicker than conventional polyethylene due to starch particles protruding as illustrated.

composition.[6] We experimented by measuring the thickness of various film products containing corn starch masterbatch with a micrometer and comparing it to the calculated average thickness per Griffin's procedure. The results were the graphs shown in Figures 5.2(a) and (b). When 1.0 mil (25 μm) film was required the extrusion line was run to produce a film measuring 1.3 mils (32.5 μm) thick with starch. As the target film thickness increased, the effects of the starch particles on measured thickness decreased.

5.3.1.3 Starch temperature limitations.   Second, corn starch was susceptible to 'caramelization' (burning) when heated above 450°F (232°C), very close to the process melt temperature of the polyethylene. Too hot a melt temperature resulted in starch caramelization, increased starch particle agglomeration, discoloration or browning of the film, an odor of burnt corn, and subsequent blow hole problems. Investigation suggested that lower melt temperature could be achieved by reducing back pressure on the screw with coarser screens than normally used to filter conventional polyethylene. We also adjusted temperature profiles on the extruder to run cooler. These actions also helped with a third starch related problem, moisture.

5.3.1.4 Starch moisture content.   Masterbatch was produced with a starch moisture level of 0.5% by weight or less. To aid in keeping moisture level down, calcium oxide was added during production of the masterbatch to act as a scavenger. High moisture levels resulted in steam formation in the polyethylene/starch film characterized by 'fish eyes', or bubbles, seen in the film. This greatly weakened film tensile strength and impact resis-

(a)

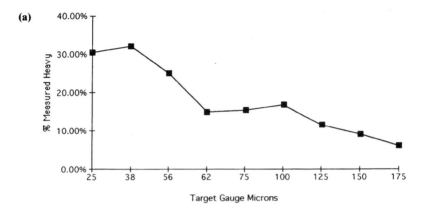

(b)

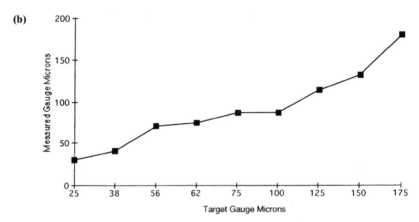

Figure 5.2 (a) The effects of starch particles on measured film thickness lessen as the film gauge increases. This graph shows the trend. For film thicker than 175 μm, the effect levels out at 5%. (b) Measured thickness versus target film thickness is shown here. This graph is used in production to tell operators how thick the PE/S film should measure to ensure enough polyethylene is present for desired physical strength.

tance. Exposure of masterbatch to ambient conditions for 24 hours or more resulted in increased moisture content by absorption and the problems just described. It was also found that melt temperature and pressure magnified moisture problems. Thus, by reducing back pressure in the screw and barrel with coarser screens, not only were melt temperature and starch caramelization reduced, but the tendency for fish eye formation was also reduced. It is believed this was because reducing back pressure resulted in a lesser pressure gradient from the extruder to the die. A lower pressure gradient would mean less tendency for flashing of residual moisture to steam at operating temperatures of 400 to 450°F (204–232°C).

5.3.1.5 Pro-oxidants One popular criticism was that PE/S films would not really degrade. Instead, after the starch was consumed by microbes, only plastic dust would remain, lasting possibly 500 years with unknown horrible effects on the environment. Also, some users of the material complained it did not degrade fast enough, for example in compost piles. Users wanted complete degradation in months, not years. Griffin and the masterbatch producers refined the corn starch masterbatch with pro-oxidant and photo-accelerator levels to enhance decomposition of the polyethylene polymer molecules regardless of starch content. The goal was to crack the polymer molecules into smaller units more readily biodegradable (i.e. consumable by microbes or other life forms). The additives, such as iron and manganese stearates, acted as free radical initiators to catalyze the oxidation of polymer molecules resulting in chain scission and gradual reduction in average molecular weight in the film matrix. These chemical reactions are temperature dependent. The reaction rates proceed faster at higher temperatures. The rate might be expected to follow the Arrhenius equation for a first order chemical reaction fairly closely.[7] The problem with this was the reaction is initiated by the blown film extrusion process temperature. The act of producing the film starts the decomposition reaction, the rate of which is dependent on temperature and oxygen, not biological activity. It was found that material made using masterbatch with high content of pro-oxidants and photoaccelerators would become useless in the box in which it was shipped in as little as six months, especially if stored in a hot (>90°F (32°C)) warehouse. By useless is meant lacking physical strength to serve its intended purpose. In some instances, customers returned product to us that could not even be picked out of the box without literally crumbling. Obviously, polyethylene can be made to degrade rapidly.

5.4 Reprocessing polyethylene/corn starch film scrap

All of the information that was learned while manufacturing PE/S blown film was useful in considering reprocessing the material. As with other products manufactured for the packaging industry, cost is always an important factor. Operating costs could be kept down if we could recycle in-plant PE/S scrap into saleable product.

5.4.1. Learning to reprocess PE/S

The first step was to run the material through conventional equipment. This equipment consisted of a 30 Hp rotary scrap grinder fitted with $\frac{3}{8}$ inch (10 mm) screens to control the size of the ground film discharged. This regrind was blown to a surge bin from which it was gravity fed to the pelletizing extruder (Figure 5.3). The extruder was an FBM Falzoni 60 mm, single screw, 30 Hp pelletiser.[8] A two-stage screw design first compressed

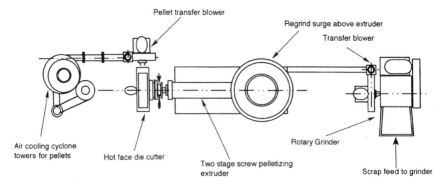

Pellet transfer blower

Regrind surge above extruder

Transfer blower

Air cooling cyclone
towers for pellets

Hot face die cutter

Rotary Grinder

Two stage screw pelletizing
extruder

Scrap feed to grinder

Figure 5.3 The FBM Falzoni 60S pelletizing machine was used to develop reprocessing parameters for PE/S film scrap.

and deaerated the regrind and the second stage metered the compressed material forward melting and mixing it. The extrudate was forced out a simple circular strand pelletizing die. A rotary die face cutter cut the emerging strands producing pellets which were then blown to a centrifugal cooling tower. After air cooling, the pellets were discharged to a collection drum. Scrap PE/S was fed to this process in the same way as conventional polyethylene and no equipment modifications were made. Initial pelletizer temperature controller settings were the same as for conventional poly-ethylene. The first attempt produced pellets that were obviously full of gas bubbles. In fact, the pellets resembled foam in appearance. They also appeared flatter than normal, an indication in the past that the extruder was running too hot. However, an attempt was made to blow film from the material anyway, since this was the first try at reprocessing. Reprocessed polyethylene was routinely blended with virgin resin at levels from 10 to 50 wt%, or run straight without virgin resin. The first step was to run the PE/S repro straight. As with the pellets, the film produced was full of fish eyes, indicating gas. Next the PE/S repro was blended with virgin resin at 10, 20 and 50 wt% to see if the gas problem persisted. It did. From experience, moisture was suspected to have been absorbed by the starch in the film. Obviously, this absorption was occurring after the film was made but before reprocessing since the gas was evident in the pellets. Also, the pelletizing process was not able to eliminate the moisture even though reprocessing temperatures were hot enough to drive off the water. A vented extruder may have helped with the problem by allowing steam to escape prior to making the pellets. However, our machine was not vented nor was such a modification readily achieved. After discussing the problem with the masterbatch producers, we decided to try adding calcium oxide (CaO) to the pelletizing process. This should act as a moisture scavenger and prevent steam formation, the same as in the masterbatch.

5.4.2   Calcium oxide moisture scavenger

Unfortunately, adding CaO to a polyethylene extruder was not easy. Powdered, dry CaO was purchased in drums and added to the extruder through a hopper feed located on the extruder barrel after the compression stage of the screw. This point had the disadvantage of acting as an outlet for air trapped in the regrind during compression. This tended to blow CaO dust all over the pelletizing operation which was both dirty and unhealthy for operators. However, the addition of 0.5 to 2 wt% CaO did eliminate the steam problem in the pellets and subsequently in the film made from them. After several production runs with this messy process, it was discovered that a calcium oxide masterbatch was available from Ampacet containing 50 wt% CaO in 20 MI polyethylene carrier resin.[9] Using this eliminated the dust and health hazards and produced the results that were sought. However, this extra step added to the cost since the CaO masterbatch was about $2.20/kg (US) and had to be added at 1 to 4 wt%.

5.4.3   Temperature control

The addition of calcium oxide to scavenge moisture solved the gas problem once the details were worked out. The solution to temperature control was more subtle but again was suggested by our experience in making PE/S blown film. In the pelletizing process, a minimum melt temperature was desirable to prevent burning the starch and to minimize the temperature dependent pro-oxidant catalyzed polymer decomposition mentioned earlier. This was achieved in two ways. First, a colder temperature profile over the length of the extruder was used and second, screens used just before the die to filter out trash and gels were changed. Finer mesh screens were used and these were changed more frequently to prevent excessive back pressure in the screw and barrel. From experience in making the film it was known that less back pressure helped lower the melt temperature. The comparative processing parameters for PE/S versus conventional polyethylene are shown in Table 5.1. These actions achieved the desired result and gave a better pellet product at a slightly higher production rate than conventional polyethylene.

5.4.4   Accounting for pro-oxidant

As mentioned before, pro-oxidant was added to corn starch masterbatch to accelerate degradation of the film for applications where this was desirable. Reprocessing film made with pro-oxidant aggravated the situation by subjecting the material to 220°C processing temperature again. Making blown film from this repro subjected it to another heating cycle. To ensure a reasonably useful life for film made from PE/S repro with pro-oxidant,

Table 5.1 Pelletizer operating parameters for reprocessing polyethylene with 6% corn starch

Parameters	Conventional polyethylene	Polyethylene and 6% corn starch
Screen pressure (Psi g)	1600–3200	1600–3200
Barrel zones	*Temperatures* (°C)	
#1 Feed	135	135
#2	140	140
#3	152	152
#4 Die	130	125
#5 Die	125	115
CaO MB (%)	0	0.5–4.0
Rate (kg hr^{-1})	75	80

Pelletizer: FBM Falzoni 60 mm, 24:1 two stage screw.
Pellets are air cooled.
Pelletizing operating parameters are comparable with those for conventional polyethylene reprocessing, with the exception of the addition of calcium oxide to act as a moisture scavanger.

reprocessing was withheld until the material was needed. Additionally, the PE/S repro was blended with some virgin materials to dilute the effects of several heat cycles, or the material was used in applications where it would have a short lag time from production to end use. If these steps were not taken, the material became useless because it had degraded too far to provide the necessary physical properties and shelf life in film made subsequently. The only alternative was disposal in a landfill.

5.4.5   Handling PE/S repro

After all these considerations and actions were taken to allow production of a usable PE/S repro, it was still possible to ruin it by allowing it to stand open too long in contact with ambient moisture which would be absorbed. This was overcome with a simple polyethylene liner for the gaylord boxes in which the PE/S repro was stored after manufacture. It was also necessary to allow the material to cool properly. Closing the box liner too early actually appeared to trap moisture that would otherwise evaporate due to residual heat left in the freshly made pellets. Closing too early also trapped too much of the residual heat. This could have been overcome by using chilled air in the pelletizing process for final pellet cooling, but was not tried. In the summer, when ambient temperatures ran near 100°F (38°C), temperatures of 140 to 150°F (60–66°C) were measured in the center of a box of newly made PE/S pellets. A 500 kg box of hot pellets retained heat for hours, and when too hot resulted in autocatalytic break down of the PE/S when pro-oxidant was present. The result was a box of fused pellets, that is a solid block weighing about

Table 5.2 Comparison of material costs

Virgin PE/S mix	Cost ($ kg⁻¹)	Cost in mix ($ kg⁻¹)
88 % Polyethylene	0.75	0.66
12 % Corn starch masterbatch	2.20	0.26
	Mix cost	0.92
Repro @ 50% of virgin		
Repro cost	0.46	0.46
2% CaO MB	2.20	0.04
Screen change added cost	0.02	0.02
	PE/S repro cost	0.52
Repro @ 70% of virgin		
Repro cost	0.64	0.64
2% CaO MB	2.20	0.04
Screen change added cost	0.02	0.02
	PE/S repro cost	0.70

Material costs for reprocessed polyethylene/starch are higher than conventional, but it is still economical to recycle in-plant PE/S scrap.

500 kg. Even after taking precautions, there were still times when gas bubbles were exhibited in the film when using PE/S repro. At those times, CaO masterbatch was added at as low a level as possible to scavenge the moisture. Usually about 1 wt% calcium oxide masterbatch was enough, although there were a few occasions when as much as 5 wt% was required. The latter situation indicated that there was either an error in the making of the PE/S repro or that an open storage container had allowed moisture absorption from the atmosphere.

5.5 Economics of in-plant recycling

A simple calculation showed that producing PE/S repro for recycling into blown film was economical even though more expensive than conventional repro. A 6% corn starch film contained 88 wt% polyethylene at $0.34/lb ($0.75/kg) and 12 wt% corn starch masterbatch at $1.00/lb ($2.20/kg). The material cost of the mix was the weighted sum or $0.42/lb ($0.92/kg). As mentioned before, reprocessing costs ran at 30 to 50% less than virgin resin cost. At 50% less, conventional repro would cost $0.21/lb ($0.46/kg). PE/S repro added costs included 2 wt% CaO masterbatch at $1.00/lb ($2.20/kg) and about $0.01/lb ($0.02/kg) for additional screen changes. This made PE/S repro material cost $0.24/lb ($0.53/kg). Even if repro-cessing costs for conventional repro were 70% of virgin resin cost, the PE/S repro would cost $0.32/lb ($0.70/kg), still less than virgin resin and 24% less than PE/S virgin mix. The economics favored recycling in-plant PE/S scrap into PE/S repro (Table 5.2).

Table 5.3 Film samples tested for physical properties

Sample	Size and thickness	BUR	Resin blend
1 Conventional polyethylene	30 × 0.0015 inch 76 cm × 38 μm	3.18	10% liner 90% LLDPE
2 Degradable repro w/6% starch	30 × 0.0015 inch 76 cm × 38 μm	3.08	8.5% liner 76.5% LLDPE 15.0% Starch MB
3 Virgin poly & starch MB, 6% starch	33 × 0.0015 inch 84 cm × 38 μm	2.62	51.0% Liner 34.0% LLDPE 15.0 Starch MB
4 Degradable repro w/6% starch	30 × 0.0015 inch 76 cm × 38 μm	3.18	100% Random color degradable repro PE/S
5 Conventional repro w/10% PE/S repro	29 × 0.0012 inch 74 cm × 30 μm	2.39	90% Conventional repro 10% PE/S repro

Five production run film samples were sent to three testing laboratories for determination of comparative physical properties. Samples were chosen so that differences in physical properties due to processing differences would be minimized.

5.6   Using PE/S repro

The processing feasibility of making PE/S repro and blowing film from it had been proved, and the economics favored using the material. However, the question of how PE/S repro film would perform still remained. Would the film and bags made from it be strong enough to perform competitively with virgin mix material? Would the PE/S repro cause premature failure due to degradation accelerated by reprocessing? It was felt that the best way to answer these questions was to test samples of film and bags made with virgin mix of conventional poly without corn starch as well as with it. It was also decided to compare samples made with PE/S repro of varying levels.

5.6.1   Comparative study of PE/S repro on film properties

'Pure' PE/S repro was easily blown into film again without the addition of virgin polyethylene or corn starch masterbatch. Starch level in the repro film was unchanged when compared with the starch level of film produced with all virgin materials.[10] PE/S repro was also blended at levels of 5 to 50 wt% with PE/S virgin mix and with conventional polyethylene repro (no added starch masterbatch). Five samples were compared using production run samples of film of approximately the same blow up ratio and thickness (Table 5.3). The physical properties of the films were tested by three separate laboratories using ASTM standard procedures.[11] Film properties

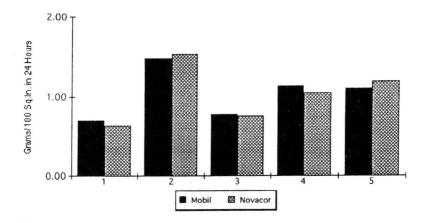

Figure 5.4 Moisture vapor transmission rate (MVTR) was comparable for all samples. Only the labs at Mobil and Novacor measured MVTR.

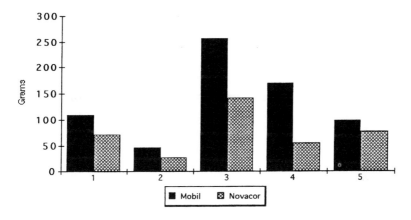

Figure 5.5 Dart impact was higher for virgin poly with starch, and lower for all the reprocessed samples, relative to the conventional polyethylene standard (sample 1).

for the samples containing PE/S repro were comparable with those for conventional reprocessed poly (see Figures 5.4 to 5.11). Some slight variations in properties between samples could be attributed to blow up ratio (BUR) differences which affect film strength. While the absolute numbers varied between the different test laboratories, the general trends between the samples were consistent. In addition to the laboratory tests run, the ultimate test was conducted by selling the product to customers. No complaints about performance of the material were received. Additionally, the practice of using PE/S repro in products where degradability is not an issue has been continued with no complaints, for the last five years (1989–93). It

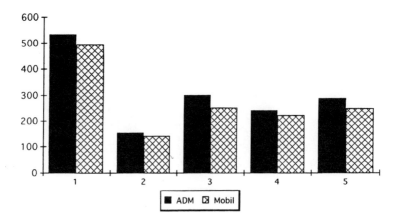

Figure 5.6 Transverse direction tensile elongation shows the effects of blow up ratio differences combined with loss of properties due to reprocessing.

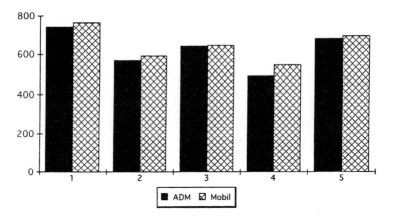

Figure 5.7 Tensile elongation in the machine direction was comparable for all samples. Novacor did not measure this characteristic.

is not known whether continued reprocessing of the PE/S material will eventually result in unusable film. However, at that point the material would have only one use anyway, as random color trash bags. If it is too weak even for that application, it is disposed in a landfill anyway. According to records kept by Manchester Packaging Co., this latter occurrence accounts for less than 5% of the PE/S material produced.

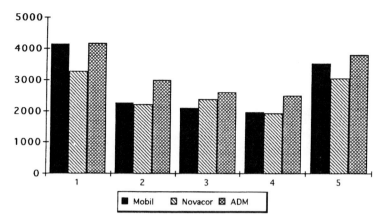

Figure 5.8 Tensile to break in the machine direction showed that starch masterbatch had a weakening effect on the film. Sample 5, which contained only 0.6% starch instead of 6.0%, showed strength comparable to conventional, even though the material was reprocessed.

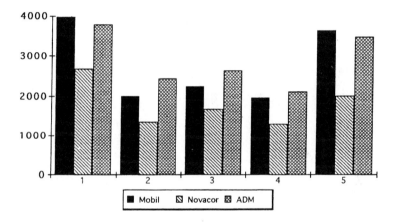

Figure 5.9 Tensile to break in the transverse direction echoes that of the machine direction.

5.6.2 Does it make sense to recycle degradable PE/S?

Experience has convinced us that recycling in-plant PE/S scrap is feasible, economical, environmentally responsible and does not interfere with recovering conventional polyethylene even if cross contamination of the two materials occurs. Cross contamination problems occur when incompatible polyethylene resins are mixed, not because of starch or degradable additives. PE/S repro cannot be blended into film for Food and Drug Administration (FDA) applications, however, since even virgin PE/S is not FDA approved. Degradable film and bags made with corn starch and pro-oxidants present a solid waste management alternative when the material

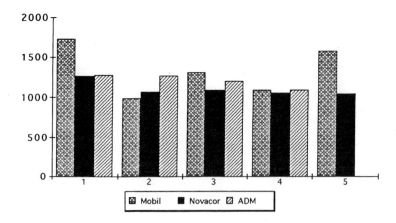

Figure 5.10 Tensile to yield in the machine direction showed again that starch masterbatch affects the strength characteristics of the film. This could be a combination of the starch particles acting as stress points and the presence of the higher melt index carrier resin used in the masterbatch.

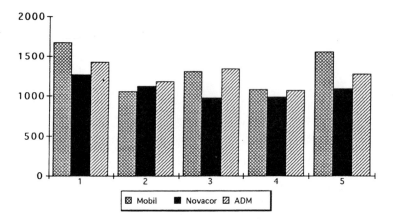

Figure 5.11 Tensile to yield in the transverse direction showed the same trends as in the machine direction. Some variation in absolute measurements between the laboratories was apparent in all the data. However, the general trends were the same.

is likely to be disposed in an environment where degradation can occur or where degradation is a desirable function of the material.

5.6.3 Recycling other degradables

Our experience with degradable plastics has been limited to polyethylene and starch blends using masterbatches from three different suppliers, all using the same basic technology. Some of what has been learned would certainly seem applicable to other degradable materials such as polyhydroxybutyrate–valerate (PHBV) made by ICI, or polyvinyl alcohol (PVOH) from Air Products and Chemicals, or polylactides now being developed by companies such as Cargill in the USA, or one of several others now in development. These polymer materials are by design readily degradable, making it impractical or even illogical to consider trying to recycle them either in the plant or post-consumer. Recycling any material requires considering whether: (1) it can be recovered from manufacturing operations in a form allowing reprocessing; (2) it is sensitive to processing conditions; (3) it will loose physical properties due to reprocessing; (4) it can be made into product straight from reprocessing or will blend with virgin material after reprocessing to make product; and (5) it can be reprocessed economically. Additionally, from an environmental point of view, recycling in-plant waste makes sense any time it can be done with less energy and without the production of more difficult waste than the production of virgin material. This area of material processing is certain to continue to develop over the coming years. Hopefully, more logic than emotion will be used in evaluating the materials' merits and limitations. We intend to continue manufacturing products in this area as long as customers indicate they want them.[12]

Notes

1. Melt index (MI) is a measure of the flowability of the plastic. A higher MI indicates lower melt viscosity at operating temperatures. ASTM standard test methods are used to measure this characteristic. For typical blown film applications 0.25 to 2.0 MI resin grades are used.
2. Starch masterbatch was originally supplied by two companies and was used interchangeably. Archer-Daniels-Midland Company (ADM), Lakeview Technical Center, 1001 Brush College Road, Decatur, Illinois 62526, supplied masterbatch under the trade name PolyClean. St. Lawrence Starch Company, Mississauga, Ontario, Canada (now Ecostar International, 181 Cooper Ave., Tonawanda, N.Y. 14150–6645), supplied masterbatch under the trade names Ecostar and Ecostar Plus. Masterbatch was also purchased from Ampacet Corporation, 250 South Terrace Ave., Mount Vernon, N.Y. 10550, under the trade name Poly-Grade. Ampacet's product was made under license with ADM.
3. Fully Compounded Plastics, Inc., 2121 South Imboden Court, Decatur, Illinois 62521–5286. Dr. J. L. Willett holds the patent for the FCP products trade named Polystarch. This material is less expensive than Polyclean or Ecostar. Several grades of masterbatch are used depending on the desired degree of degradability. FCP supplies a masterbatch with starch only, no pro-oxidant, called Polystarch-N; a compost masterbatch with pro-oxidant and photoinitiator for faster degradation, called Polystarch-Plus; and a mulch grade with higher levels of pro-oxidant and photoinitiator, called Polystarch-M. Other masterbatch producers offer similar grades.
4. Sterling Extruder Company, 901 Durham Ave., South Plainfield, N.J. 07080, is now part of Davis-Standard, Inc., 38 Brunswick Ave., Edison, N.J. 08817. Gloucester Engineering Company, now Battenfeld/Gloucester Engineering Co. Inc., Blackburn Industrial Park, Gloucester, Ma. 01930.
5. Sano, Inc. is now a Division of Cincinnati Milacron, Cincinnati, Ohio.
6. Polyethylene density varies with resin grade, but for general use low density and linear low density polyethylene have a density of 0.92 g cm^{-3}. Corn starch density is 1.49 g cm^{-3}, but can show much lower values when intensely dried because of internal void

development. Dr. G. J. L. Griffin explained the effects of the starch on average density of the polyethylene film blown using masterbatch in a paper titled *Mechanical properties of starch/polyethylene films*, a private communication.

7. Selling degradable products has a certain risk because the material will slowly deteriorate with time since the pro-oxidant, if present, is activated by the process of extrusion at 220°C. Our experience has led us to label every case and invoice with a warning that the material has a shelf life which is affected by storage temperature. Working with Dr. J. L. Willett of Fully Compounded Plastics, we developed a simple and conservative way to predict shelf life for our customers. The derivation is shown below.

Estimation of shelf life from thermal ageing data per Dr. J. L. Willett I I, FCP, Inc. 08/23/91 CRA.

(a) Assuming an Arrhenius (temperature activated (rate)) process for the thermal degradation of polyethylene film with corn starch, pro-oxidant, and photoaccelerator additives. This applies to FCPs Polystarch-Plus compost grade masterbatch.

$$k(T) = k(r)\, e^{(-E/RT)}$$

where $k(T)$ = rate at temperature $T°K$

$$k(T_2) = k(T1)\, e^{[(-E/R)*\,(1/T_1 - 1/T_2)]}$$

R = gas constant; E = activation energy

(b)   Lifetime $t(T) = A/k(T)$

where A = constant, $t(T)$ = days of life at temperature T.

$$1/t(T_2) = [1/t(T_1)]\, e^{[(E/R)*\,(1/T_1 - 1/T_2)]}$$
$$[t(T_2)/t(T_1)] = e^{[(E/R)*\,(1/T_1 - 1/T_2)]}$$

This is the equation of storage life depending on temperature, where T_1 and T_2 are two temperatures for relative shelf life.

(c) Inserting experimentally determined values into the equation for predictions: T_1= 60°C (333°K), T_2 = 25°C (298°K), R = 1.987 cal mol^{-1} °K^{-1}, E = 16000 cal mol^{-1}, (from FCP & Ecostar data per Dr. J. L. Willett) $t(T_1)$ = 30 days from oven studies (J. L. Willett)

$$t(T_2)/t(T_1) = e^{[(16\,000/2)*\,(1/298 - 1/333)]} = 17.12$$

This is interpreted to mean the degradation reaction for material stored at 60°C will proceed 17 times faster than material stored at 25°C. Material stored at 60°C had 0% elongation at 30 days. Thus material stored at 25°C would have an expected storage life of 17 × 30 = 510 days.

Table of predicted shelf life depending on storage temperature

Reference temperature $T_1(°K)$	Comparative storage temperature T_2 (°K)	Estimated Rate $t(T_1)/t(T_2)$	Life in days at T_2
298	298	1.00	514
298	303	1.56	329
298	308	2.40	214
298	313	3.65	141
298	323	8.10	63
298	333	17.12	30

The last row in this table was determined by baking film samples in an oven and measuring physical properties. The results showed 0% elongation in the sample after

30 days at 60°C. The other numbers in the table were calculated from the above equations using this data from the oven study by Dr. Willett.

8. FBM Falzoni, 440122 Bondeno, Italy, manufactures the reprocessing equipment used to develop the information in this chapter. It is sold in the USA by Plastic Systems, Inc., Algonquin, Illinois.

9. Ampacet CaO masterbatch, product #10967, is 50% loaded CaO in 20 MI polyethylene carrier resin. A similar product is available from Fully Compounded Plastics called Aquanil. Both products are blended with PE/S as a moisture scavenger.

10. Film starch level was tested by Archer-Daniels-Midland Company using FTIR spectroscopic analysis.

11. Samples of all five polyethylene and polyethylene/starch films were pulled from normal production runs in March 1990. These samples were submitted to three unrelated companies for physical properties analysis. Requested information included moisture vapor transmission rate, dart impact, tensile elongation, tensile yield and tensile break. The three companies were not given any information regarding the samples so the tests were done blind. The results presented in Figures 5.4 to 5.11 were determined by:

Archer-Daniels-Midland Company, Lakeview Technical Center, Decatur, Illinois, Dr. George Poppe.
Mobil Chemical Company Research and Development, Edison, New Jersey, Frank J. Velisek.
Novacor Chemicals, Inc., Technical Services, Schaumburg, Illinois, Chris Pajak.

ASTM methods and procedures were used to make the physical property measurements.

12. Manchester Packaging Company is a custom manufacturer of polyethylene film and bags. The company is located at 2000 East James Blvd., St. James, Missouri 65559. MPC has been involved with degradables using corn starch since 1987. The company was awarded $10 000 by The National Corn Growers Association in 1990 at the 32nd Annual NCGA Corn Classic, held in Phoenix, Arizona, for work in the development of polyethylene/corn starch film products. MPC continues to manufacture polyethylene/corn starch film and bags for degradable uses such as municipal composting programs. PE/S film and bags are also made for non-degradable applications, to take advantage of the special properties associated with the product. MPC markets these products under the trade name D-grad®.

References

1. Rathje, W. L. (1991) Once and future landfills, *National Geographic*, **179** (5), 116–134.
2. Griffin, G. J. L. (1977) *US Patent* 4 021 388 and 4 016 117.
3. Wilder, R. V. (1989) Degradability 1: 'Disappearing' package: pipe dream or savior? *Modern Plastics*, 40–45.

6 Test methods and standards for biodegradable plastics

K. J. SEAL

6.1 Introduction

Polymers and plastics which undergo, or are claimed to undergo, some form of physical or chemical change as the direct result of biological activity have established themselves in a variety of niche markets in the packaging, horticultural and medical sectors. In order that claims of biodegradability can be supported for marketing and regulatory purposes, standard test methods acceptable at national and preferably international level must be introduced. Regulatory and consumer pressure demand increasingly that the goods we use do not cause an unacceptable risk to the quality of the environment when we discard them. The assessment of biodegradability is one such factor which contributes towards the risk assessment process now becoming part of the requirement for chemicals and preparations which find their way ultimately into the environment.

The testing of chemicals for their biodegradability has been carried out for over 30 years and protocols are enshrined in data requirements for notification and registration of new chemical substances in Europe, the USA and many other developed countries. The methodology was designed for relatively simple water soluble chemical products such as surfactants where a ranking of biodegradability could be related to chemical structure using short term tests (28 days duration). The regulators have included biodegradability assessments as a first tier requirement for all chemicals and preparations which are produced in annual quantities of greater than one tonne. The consumer perceives biodegradability as a desirable property and this has been used for political ends. Unfortunately, perception and fact do not often go hand in hand and the development of scientifically acceptable methodologies classically lags behind the institution of politically expedient measures. The evaluation of polymers and plastics for their biodegradability is one area where the situation described above is particularly acute. Current standard methodology is either totally inadequate to describe biodegradability, or does not provide meaningful practically related data on which to make an environmental assessment.

Polymers used in defined situations such as the human body for wound repair can be reasonably easily assessed for their degradative properties: whether this is termed biodegradation is a matter for discussion outside this chapter.

6.2 Defining biodegradability

Biodegradation is the natural process by which organic chemicals in the environment are converted to simpler compounds, mineralised and redistributed through elemental cycles such as the carbon, nitrogen and sulphur cycles. Biodegradation can only occur within the biosphere and microorganisms play a central role in the biodegradation process. The total mineralisation implied in this definition represents an ideal situation which in practice probably never occurs in the majority of natural polymeric materials. It is desirable for lignocellulosic wastes to retain some recalcitrant material which is then incorporated into the matrix of the soil improving its structure. It is thus dangerous to talk of achieving a standard which specifies 100% biodegradability in a material, as such a standard cannot be achieved let alone be quantitatively determined. However, there must be definitions against which test methods can be developed and acceptance criteria be judged.

A number of standards authorities have sought to produce definitions for biodegradable plastics and some of these are reproduced below.

ISO 472: 1988—*A plastic designed to undergo a significant change in its chemical structure under specific environmental conditions resulting in a loss of some properties that may vary as measured by standard test methods appropriate to the plastic and the application in a period of time that determines its classification. The change in chemical structure results from the action of naturally occurring microorganisms.*

ASTM sub-committee D20.96 proposal—*Degradable plastics are plastic materials that undergo bond scission in the backbone of a polymer through chemical, biological and/or physical forces in the environment at a rate which leads to fragmentation or disintegration of the plastics.*

Japanese Biodegradable Plastics Society [1] draft proposal—*Biodegradable plastics are polymeric materials which are changed into lower molecular weight compounds where at least one step in the degradation process is through metabolism in the presence of naturally occurring organisms.*

DIN 103.2 working group on biodegradable polymers—*Biodegradation of a plastic material is a process leading to a change in its chemical structure caused by biological activity leading to naturally occurring metabolic end products.*

A plastic material is called biodegradable if all its organic components undergo a total biodegradation. Environmental conditions and rates of biodegradation are to be determined in standardised test systems.

A plastic material is defined as the polymer plus additives plus processing.

The definitions produced above and others in the literature are pragmatic in their approach by addressing degradation from the viewpoint of the plastic as a material and not assessing its ultimate fate in the environment. Additionally they do not indicate any criteria for a practical classification of biodegradability by relating the extent to which a plastic has biodegraded within a chosen timeframe.

In order to address the environmental aspects the extra dimension of complete or ultimate biodegradation will be emphasised in this chapter as the main criterion for defining a biodegradable polymer.

6.3 Criteria used in the evaluation of biodegradable polymers

A large number of intrinsic properties of a polymer can be measured and these will be familiar to polymer chemists and engineers. Some properties whilst producing absolute indicators that the polymer has undergone scission, e.g. molecular weight distribution or intrinsic viscosity changes, require specialist analytical equipment but tell us little of the ultimate mineralisation or biodegradability of the polymer.

Weight loss has limited use particularly where the polymer fragments and the integrity of the specimens are lost during the test. Mechanical property changes are difficult to interpret in relation to structural alterations although they may be sensitive to small changes in molecular weight.

A simple visual examination for the presence of actively growing microorganisms on specimens in a nutrient medium and their surface effects [2] where the polymer is the sole carbon source can be a useful first stage qualitative indicator that biodegradation is occuring. Sophisticated tools such as Fourier transform infra red (FTIR) and nuclear magnetic resonance (NMR) are normally reserved for research purposes.

Biodegradability test protocols adopted by the Organisation for Economic Cooperation and Development (OECD)[3] have used specific analyses to follow the disappearance of the parent chemical during the exposure period and more latterly the detection and measurement of oxygen consumption and/or CO_2 evolution which directly indicate microbial metabolism of the substrate. CO_2 in particular is considered a good criterion for mineralisation and hence ultimate biodegradation of an organic molecule. Specific analysis is limited in that it indicates only primary biodegradation of the molecule and does not, unless the complete degradative pathway is analysed, give any clue to the possible accumulation

Table 6.1 Assessment of biodegradation of chemical substances using the half-life classification scheme of Howard *et al*.[4]

Biodegradation rate	Half-life ranges	
	Lower	Upper
Fast	1 day	7 days
Moderately fast	7 days	4 weeks
Slow	4 weeks	6 months
Resistant	6 months	12 months

of intermediate products. Indeed if specific analysis is used as a measure of biodegradability it can easily lead to claims of 100% biodegradability where none of the original test chemical is detectable in the test system on completion of the exposure period!

This author believes that testing strategies must in the first instance include ultimate biodegradation or mineralisation as the criterion in assessing a biodegradable polymer. In this way there is no need for specific analysis and comparison between test and reference materials is possible. A number of standards committees, especially ASTM sub-committee 20.96 and CEN TC261, are considering methods employing this criterion. There is much work still to be done in determining an acceptable level of biodegradation which must be achieved within a certain timeframe and whether the classification used by the OECD into ready and inherently biodegradable substances (see section 6.4) is relevant to polymers.

The calculation of half-lives from the respiration data has been used to compare biodegradability for organic chemicals [4].

Differentiating biodegradation rates into four classes from fast to resistant (Table 6.1) was based upon the half-life data ranges obtained from screening tests. Such an approach could be adapted for biodegradable polymers with some modification to the time values.

6.4 Tiered systems for evaluating biodegradability

The assessment of chemicals for their effects on and fate in the environment, which includes biodegradability, involves a tiered approach. Regulatory authorities have adopted this system to limit the amount of testing, and hence cost, to that required for a chemical to reach an acceptable result upon which a risk assessment may be made. Testing starts at the first tier and progresses through increasing levels of environmental simulation. When the chemical meets the required acceptance level of biodegradability or when sufficient data on its biodegradation profile is obtained the testing stops. In practice the quantity of chemical manufactured annually will also trigger extra tiers of testing.

The first tier consists of short term screening tests under stringent conditions run in a controlled environment in the laboratory. These tests provide limited opportunity for biodegradation and acclimatisation to the test substance to occur.

Such chemicals which reach the pass level in these tests are assumed to be able to biodegrade rapidly in the environment and may be classified as *readily biodegradable.*

The second tier involves tests which allow prolonged exposure of the test substance to a natural inoculum of microorganisms. A more favourable ratio of microbial inoculum to the test substance is provided, and there is an acclimatisation period before testing commences. A substance which gives a positive result in a test of this type may be classified as *inherently biodegradable.* Such a classification means that because of the favourable conditions employed in the test, rapid and reliable biodegradation of the test substance may not be assumed.

The third tier of testing is designed to simulate repeated application of a test substance to a simulated disposal route. Model rigs to simulate sewage treatment have been validated and accepted for testing at this tier. Other protocols may be custom designed to suit particular applications. The tests are necessarily long term and often require identification of intermediates to assess likely accumulation products. High annual production levels of the substance may trigger this tier. Because of their cost tests of this type require careful validation and justification.

6.5 Choice of environment

To produce meaningful data capable of interpretation for the variety of disposal routes for biodegradable polymers the tests must, even at the screening level, go some way to incorporating salient features of the particular environmental compartment into the experimental design. The most likely disposal routes are shown in Table 6.2.

Test systems for some of these compartments are better characterised than for others. For example, the aqueous aerobic situation is well served for test protocols whilst anaerobic tests are still at a development stage,

Table 6.2 Disposal routes for biodegradable polymers

Terrestrial	Aqueous
Landfill	Groundwater
Soil	Waterways
Composting	Marine
	Sewage
	Sediments

particularly at the screening level. Testing in sediments, soil and compost require custom-designed approaches which still lack standardisation.

Polymeric materials are likely to find their way into both the aqueous and soil compartments of the environment. In particular the potential for biodegradable polymers in disposable products means that the sewage and marine environments must also be considered. Such situations can be incorporated into test protocols even at the first tier stage of testing by the use of suitable test inocula and the appropriate physiological conditions.

The biodegradability of polymers under anaerobic conditions, such as those encountered in an anaerobic sewage digester, is an area which is poorly understood and characterised. Biodegradability rates may be many times less under such conditions, as oxidative reactions involving oxygenases and molecular oxygen cannot take place in the highly reduced atmosphere. Test methods in this area rely on the simulation of an anaerobic digester in the laboratory and the comparative measurements of CH_4 and CO_2 between a test rig and a control. The extent to which the material has biodegraded is calculated by comparison of the total carbon released as gas with the theoretical maximum derived from the molecular formula.

The composting environment increasingly represents a potentially significant route of disposal with the revival of the composting process in Europe and the USA for municipal waste treatment. Advances in waste separation technology, legislation demanding that countries reduce total amounts of waste and increased consumer willingness to separate waste streams at source have encouraged authorities to revisit composting as a method for waste minimalisation which is also an 'environmentally friendly' technology. Composting is an ideal environment for the rapid degradation of organic matter. The high temperatures which characterise an efficient composting process can also be employed to increase the rate at which chemical initiators start the degradation process in some degradable polymers specially designed to respond to composting conditions.

Composting is not an easy process to 'miniaturise' or standardise for screening purposes in the laboratory. Model composting units have been designed but they tend to be smaller versions of full scale plants. Because of the heterogenous nature of composts it is difficult to run controlled experiments and physical and mechanical assessments of samples placed in the compost will have to be combined with data from other aerobic tests to gain the whole biodegradability pattern.

6.6 Choosing the most appropriate methodology

The use of decision trees is gaining popularity in following a programme of testing in a logical fashion. This approach rules out the generation of unnecessary and irrelevant data. An example decision tree is presented in

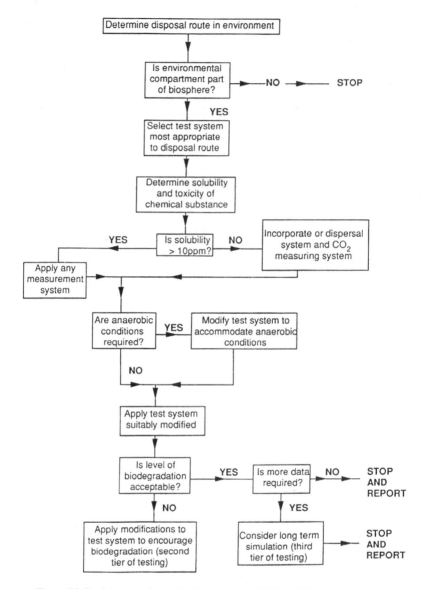

Figure 6.1 Decision tree for evaluating biodegradability of chemical substances.

Figure 6.1. At any decision point the assessor is only faced with two alternative routes to follow which will be based upon the prior acquisition and interpretation of test data.

The first decision requires data on the concentration released and the compartment of the environment to which the product is released. If there is no likelihood that the polymer will ever come into contact with the biosphere then it is senseless carrying out a biodegradability assessment. This is of course a hypothetical situation for biodegradable polymers as disposal to the biosphere is highly likely and one or more routes of exposure must be defined. From this a suitable methodology can be chosen for the first tier of testing.

If the polymer 'passes' this first tier then no further testing is necessary. 'Failing' the test triggers the decision to proceed to the next tier. Decisions are thus made in a stepwise fashion until the hypothesis that the polymer is biodegradable has been proved or refuted.

Having demonstrated the inherent or intrinsic biodegradability of the polymer it may be desirable to proceed to an applications related testing programme. Here the polymer is fabricated into its final form and exposed to simulated disposal routes. Information useful for marketing purposes can be gained by assessing changes in mechanical strength, or the time it takes for fragmentation to occur and thus the litter problem to disappear. This information is of practical value and complements the fundamental data gained from the inherent tests.

6.7   Description of current test methods

6.7.1   Screening tests for ready biodegradability

The five test protocols described in the OECD guidelines for testing of chemicals [3] have been accepted by the EC and the EPA. These are listed in Table 6.3.

All of these tests are similar in a number of respects: they are aquatic and operate under aerobic conditions, the test substance provides the sole source of carbon, it is exposed to a relatively low level of inoculum, and a non-specific method of analysis is used to follow the course of biodegradation. All tests run for 28 days without any acclimatisation period.

Three non-specific analyses are employed: dissolved organic carbon (DOC), dissolved oxygen concentration and CO_2 evolution (Table 6.4). Those methods where DOC is measured require the test substance to be soluble in water at the concentration used in the test. This rules out 301A and 301E for use with biodegradable polymers. Dispersion of the insoluble

Table 6.3   Five test protocols described by OECD for ready biodegradability

Title	OECD reference
Modified AFNOR test	301A
Modified Sturm test	301B
Modified MITI test	301C
Closed bottle test	301D
Modified OECD screening test	301E

Table 6.4 Pass level criteria for ready biodegradability tests

Test	Test parameter	Criteria for pass level
Modified AFNOR	DOC loss	$\geqslant$ 70% loss of DOC
Modified Sturm	CO_2 evolution	$\geqslant$ 60% yield of CO_2
Modified MITI	O_2 consumption or loss of parent compound	$\geqslant$ 60% of BOD $\geqslant$ 70% loss of parent compound
Closed bottle	O_2 consumption	$\geqslant$ 60% of BOD
Modified OECD screening	DOC loss	$\geqslant$ 70% loss of DOC

Note: Pass levels are percentage of theoretical total DOC, O_2 consumption or CO_2 evolution for test substance.

substance is important to maximise contact with the planktonic inoculum and this is normally achieved by stirring the medium.

The modified Sturm test seems to be the preferred technique for polymeric materials. It has been specified by the Italian authorities for assessing biodegradable polymers and is currently being evaluated by the Biodegradable Plastics Group of the International Biodeterioration Research Group [5]. A similar ASTM method (D5209) for plastic materials was at the approval stage in 1992.

The principle of the Sturm method is as follows. To a chemically defined mineral nutrient solution free of organic carbon the test substance is added at two concentrations (10 and 20 mg l^{-1}).

An inoculum of sewage microorganisms is added (1–20×10^6 ml^{-1}) to the solution. The test systems with suitable controls are incubated at ambient temperature with stirring for 28 days. The CO_2 evolved is trapped

Test vessels

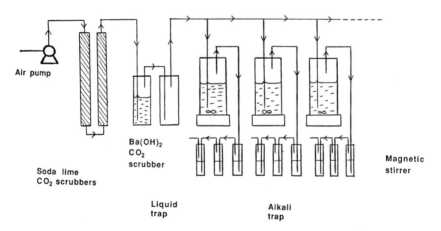

Figure 6.2 Schematic arrangement of modified Sturm test.

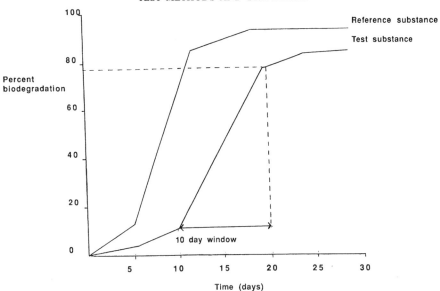

Figure 6.3 Graph to show the performance required of a readily biodegradable test substance.

in alkali and measured as carbonate by either titration or the use of a carbon analyser. The test arrangement is shown in Figure 6.2.

After analysis of the data with respect to suitable blank controls the total amount of CO_2 produced by the chemical over the test period is determined and calculated as that percentage of the total CO_2 which the chemical could have theoretically produced based upon its total carbon content. In this way, because a proportion of the carbon will be incorporated into biomass, the total CO_2 and hence biodegradation levels can never reach 100%. With this in mind, more realistic levels have been recommended.

For a chemical substance to be regarded as readily biodegradable it should produce greater than 60% of its theoretical total within 28 days. This level should be reached within 10 days of the biodegradation reaching 10% (Figure 6.3).

The reproducibility of the test is quoted in the guidelines [3] as being ± 5% but this is based upon work with soluble materials. For insoluble materials the form in which they are presented to the test system and the efficiency of dispersion will affect the reproducibility. The use of a powder or teased fibres will give the best surface to volume ratio but care should be taken to ensure that clumping or adsorption to the walls of the test vessel are avoided.

An alternative approach is to determine the biochemical oxygen demand of the polymer over 28 days (BOD_{28}). This method is probably the most stringent of the aqueous screening tests because of the low level of inoculum (in the order of 10^2 microorganisms ml^{-1}) and the limited amount of test substance which can be added (normally between 2 and 4 mg l^{-1}). The calculation in arriving at the correct application level is based on the theoretical oxygen demand (determined by calculation) or chemical oxygen demand (determined experimentally) of the test substance being not more than one half of the maximum dissolved oxygen level in water at the temperature of the test. This is determined as follows.

Taking glucose as an example, the theoretical or chemical oxygen demand is calculated as 1.07 mg O_2 mg^{-1} glucose and the concentration of dissolved O_2 in water at 20°C is about 9 mg l^{-1}. To ensure that 50% of the dissolved O_2 remains at the end of the test period the total oxygen demand must not exceed 4.5 mg l^{-1}, thus the maximum concentration of glucose used should be 4.2 mg l^{-1}.

The theoretical oxygen demand (TOD) of a chemical with formula $C_cH_hO_o$ is as follows: TOD $= 16[2c + h/_2 - o]$/molecular weight.

Determination of the theoretical oxygen demand requires a knowledge of the formula of the test substance. This may be difficult to determine accurately for polymeric materials and thus a chemical oxygen demand will need to be carried out.

The principle of the closed bottle test method is as follows. A predetermined amount of the test substance is added to a chemically defined mineral salts solution. The solution is inoculated with sewage microorganisms and then dispensed into closed bottles.

The bottles are incubated in the dark at 20±1°C and periodically assessed for their dissolved oxygen content. The oxygen demand is calculated and compared with the theoretical or chemical oxygen demand of the test substance.

Polymers would have to be prepared as finely divided powders and their continuous dispersion in the nutrient solution assured. This can only effectively be done using magnetic stirrers and this may preclude the use of this test as for one test substance at least 25 bottles are used.

6.7.2 Tests for inherent biodegradability

A failure to reach an acceptable pass level in the screening tests will normally necessitate the next tier of testing. There are four inherent test protocols recommended by the OECD which are listed in Table 6.5.

Table 6.5 Four test protocols recommended by OECD for inherent biodegradability

Title	OECD reference
Modified SCAS test	302A
Modified Zahn-Wellens test	302B
Modified MITI test	302C
Inherent biodegradability in soil	304A

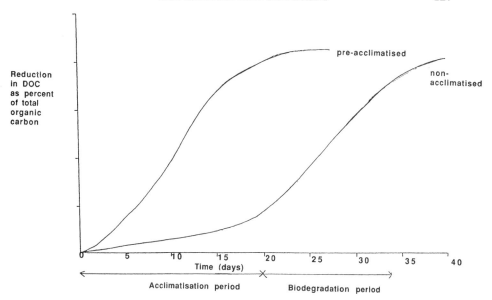

Figure 6.4 Comparison of biodegradation rates from an inherently biodegradable test substance using preacclimatised and non-acclimatised sewage inoculum and the Zahn-Wellens test.

As these methods test the inherent or intrinsic ability of the test substance to biodegrade, there are no pass or fail criteria. However, at least 20% biodegradation is recommended by the OECD to indicate inherent biodegradability. The protocols allow the investigator greater flexibility to determine the profile of biodegradation in a more favourable environment. Although 28 days is the suggested exposure period, a longer time course is permissible and prior acclimatisation procedures such as elective enrichment can be used to select a microbial consortia able collectively to metabolise the test substance (Figure 6.4). This is particularly important for xenobiotic compounds where degradative capability resides in a small part of the microbial community and metabolism pathways need time to be induced.

The use of radiolabelled test substances is of benefit in this tier of testing, particularly where low degradation rates are anticipated and an unequivocal measurement of mineralisation ($^{14}CO_2$ from ^{14}C labelled test substance) is required in the presence of any endogenous respiration. Toxic compounds only require the use of radiolabelled substrates where sub-lethal concentrations below the resolution of the normal detection system have to be used. Radiolabelling techniques are extremely sensitive and unequivocal in the result they produce. They are also valuable particularly for insoluble materials such as polymers where a carbon balance is required to account for the fate of all the carbon in the test system. The labelled polymer bound up with the microbial sludge can be differentiated

Table 6.6 Test substance addition levels for inherent biodegradability tests

Soil type	Test substance levels
Modified SCAS	20 mg l⁻¹ as DOC
Modified Zahn-Wellens	50–400 mg l⁻¹ as DOC
	100–1000 mg l⁻¹ as COD
Modified MITI	30 mg l⁻¹ test substance
Inherent biodegradability in soil	10 mg kg⁻¹ soil or radioactivity level of 37–185 KBq (= 1–5 μCi 100 μl⁻¹) for ¹⁴C labelled substances

Table 6.7 Soils recommended by OECD inherent test

Soil type	Characteristics	
Alfisol	pH:	5.5–6.5
	Organic carbon:	1–1.5%
	Clay content:	10–20%
	Cation exchange capacity:	10–20 meq
Spodosol	pH:	4.0–5.0
	Organic carbon:	1.5–3.5%
	Clay content:	≤ 10%
	Cation exchange capacity:	< 10 meq
Entisol	pH:	6.6–8.0
	Organic carbon:	1–4%
	Clay content:	11–25%
	Cation exchange capacity:	> 10 meq

by burning and trapping the labelled CO_2. However, they are expensive, require careful consideration of controls and need specialist handling facilities.

The test systems vary in design but they measure either O_2 consumption, CO_2 evolution or loss in DOC. For the aqueous systems, the inoculum is normally sewage derived but can be a mixture from environmental sources (groundwater and soil). It is not permitted to use selected specific cultures of microorganisms not derived from the acclimatisation process. Table 6.6 shows the levels of test substance to be added. Whereas for the Zahn-Wellens and MITI tests the inoculum is added only once, the SCAS test attempts to simulate a semi-continuous treatment system by using a 'draw and fill' technique for daily addition of fresh inoculum and test substance. For the soil test, soils representing different types are used. The minimum requirement is normally one of high organic matter and one of low organic matter. The former, a clayey loamy type, and the latter, a sandy loam. OECD specify three types (Table 6.7) to include an acid soil. As biodegradability is the parameter to be investigated, then the biomass activity and

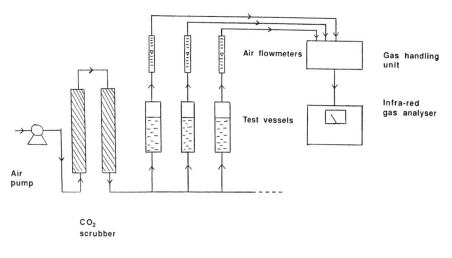

Figure 6.5 Schematic arrangement for assessing biodegradability in soil using infrared detection of CO_2 evolution.

composition is important. Biomass activity is generally directly proportional to organic matter levels in soil.

As for the screening tests, the best means of applying the test substance, if a polymer, is as a powder. Correct incorporation into soil is important and preparation of the soils by sieving and standardising on moisture content is a necessary prerequisite to optimising the conditions for the test.

As the only way to assess non-specifically the biodegradability of polymers is by monitoring CO_2 production, the methodology for the aqueous systems must be modified to simulate the Sturm test design. The most appropriate candidate for this modification is the Zahn–Wellens test. However, it is important that for long term studies the integrity of the 'plumbing' is maintained to prevent any ingress or loss of CO_2. This practical problem has been experienced by the author.

A method used in the author's laboratories for long term studies involves the continuous or semi-continuous monitoring of CO_2 using a flow through system and detection by infrared analysis. Up to 24 'bioreactors' can be sequentially monitored for CO_2 production and both rates of production and total production can be calculated. The method is particularly applicable to soils and incorporates controls. Unlike the titration method, the test system is never opened to remove and replace the trapping solutions which can affect quality of the data. A scheme for a continuous flow system is shown in Figure 6.5.

6.7.3   Tests for simulation studies

Simulation studies range from laboratory designed equipment, which repli-
cates aerobic sewage treatment and anaerobic sludge digestion, through
to exposure trials where material is buried in soil or submerged in activated
sludge, freshwater or marine environments. Exposure trials require that
the samples be securely held on some form of racking for aqueous environments. The racks, normally made of stainless steel, are submerged in the
test situation and samples periodically removed. Here it is not possible to
assess ultimate biodegradability and such tests are normally an extension
of the tests described in section 6.8.

Laboratory simulations can be used to determine ultimate biodegradation but their real value is in assessing long term effects on disposal
systems as a result of continuous dosing. The OECD Coupled Units test
(OECD 303A) simulates an activated sludge sewage treatment system but
its application for polymers would be difficult as DOC is the parameter
used to assess biodegradability.

6.8   Other methods for assessing polymer biodegradability

Several standard methods have been developed over the past thirty years
in order to assess the resistance of plastics to microbiological degradation.
They are worthy of mention, partly because they have been used to assess
biodegradable polymers, and partly because they should be considered
where in-service resistance data is required. They may also be used to
assess pragmatic criteria for defining biodegradable polymers such as fragmentation and disintegration in the environment. There are three test
systems commonly used; the petri dish screen, the environmental chamber
and the soil burial.

6.8.1   Petri dish screen

This test is used in USA (ASTM), German (DIN), French (AFNOR),
Swiss (SN) and International (ISO) standards (Table 6.8). The principle
of the method involves placing the test material (2.5×2.5 cm^2) on the
surface of a mineral salts agar in a petri dish containing no additional
carbon source.

The test material and agar surface are sprayed or painted with a standardised mixed inoculum of known fungi or bacteria (Table 6.9). The petri
dishes are sealed and incubated at a constant temperature for between 21
and 28 days. The test material is then examined for the amount of growth
on its surface and a rating given (Table 6.10). The more growth on the

Table 6.8 Standards used for resistance testing of plastics

Title	Standards authority and number
Plastics: Determination of behaviour under the action of fungi and bacteria	ISO 846 NFX41–514
Basic environmental testing procedures for electronic equipment — Test J, Mould growth	BS 2011 Part 2.1 Test J
Resistance of plasticisers to attack by microorganisms	NFX41–513
Determination of the resistance of plastics to fungi and bacteria	DIN 53 739
Standard practice for determining the resistance of synthetic polymeric materials to fungi	ASTM G21–70
Standard practice for determining the resistance of synthetic polymeric materials to bacteria	ASTM G22–76
Standard practice for determining the resistance of synthetic polymeric materials to algae	ASTM G29–75

Note: BS = British Standards Institute, DIN = German Standards Institute, ISO = International Standards Organisation, NFX = French Standards Institute.

Table 6.9 Test strains of fungi and bacteria used for resistance testing of plastics

Test strain	Culture collection number	Standard
Aspergillus niger	IMI 17454	BS 2011 Part 2.1 J
	IMI 45551	ISO 846
	IMI 91855	ASTM G21–70
Aspergillus terreus	IMI 45543	BS 2011 Part 2.1J
Aureobasidium pullulans	IMI 45553	BS 2011 Part 2.1J
	IMI 145194	ISO 846
Chaetomium globosum	IMI 45550	ASTM G21–70
		ISO 846
Paecilomyces variotii	IMI 108007	BS 2011 Part 2.1J
		ISO 846
Penicillium funiculosum	IMI 114933	BS 2011 Part 2.1J
		ISO 846
Penicillium ochrochloron	IMI 61271i	BS 2011 Part 2.1J
Scopulariopsis brevicaulis	IMI 49528	BS 2011 Part 2.1J
Trichoderma viride	IMI 45553i	BS 2011 Part 2.1J
		ASTM G21–70
		ISO 846

surface, the more likely is the material intrinsically able to support growth and thus the more likelihood that it will fail inservice.

Weight loss, mechanical or electrical tests can be carried out on the test materials after exposure provided that the correct types of specimens (e.g. dumbbells) have been used in the test. The validity of this type of test and the use of visual assessment alone has been questioned by Seal and Pantke [6] for all plastics. They recommended that mechanical properties should be assessed to support visual assessments. Such tests must be treated with caution when extrapolating the data to field situations.

Table 6.10   Rating scheme based on visual assessment used by ISO 846 for assessing fungal resistance of plastics

Visual assessment	Rating	Evaluation
No growth apparent even under the microscope	0	The material is not a nutritive medium for microorganisms
Growth invisible or hardly visible to the naked eye but clearly visible under the microscope	1	The material contains nutritive substances
Slight growth covering less than 25% of the specimen surface	2	The material is not resistant to fungal attack and contains nutritive substances
Growth covering more than 25% of the specimen surface	3	As for rating 2

6.8.2   Environmental chamber method

Environmental chambers employ high humidity (> 90%) situations to encourage microbial (in particular fungal) growth. Strips or prefabricated components of the test material are hung in the chamber, sprayed with a standard mixed inoculum of known fungi (Table 6.9) in the absence of additional nutrients and incubated for 28 to 56 days at constant temperature. A visual assessment is subsequently made and a rating given based on the amount of growth on the material (Table 6.10). This test is particularly stringent and was designed to simulate the effects of high humidity conditions on electronic components and electrical equipment. Growth of fungi across a printed circuit board can result in a gross systems failure in a computer system or military equipment. Such a test system is valuable in assessing how biodegradable polymers will perform under such conditions whilst in service.

6.8.3   Soil burial tests

Tests based upon this methodology evaluate in-service soil contact exposure.

The material is buried in soil beds prepared in the laboratory using standard sieved soil; often a commercial soil based compost such as John Innes number 1 or number 2 is specified. The soil beds are normally conditioned for up to 4 weeks prior to use and may be supplemented with organic fertiliser to encourage an active microbial flora. The microbial activity is tested using a cotton textile strip which should lose 90% of its tensile strength within 10 days of exposure to the soil. Currently no other reference materials are recommended, although for plastic materials a standard alternative able to demonstrate the degradative capabilities of the microbial flora with respect to plastic should be sought. The soil beds containing the samples are incubated at a constant temperature for between 28 days and 12 months. The moisture content is normally set at

20–30%, although it is better calculated as a percentage (40–50%) of the soil's maximum water holding capacity. This then accounts for different soil structures and ensures that the soil does not become unduly wetted or is too dry for optimal microbial activity. Samples are removed for assessment of changes in their properties such as weight loss, mechanical strength changes or a microscopic (light and scanning electron microscopy (SEM)) examination to assess surface damage and to look for the presence and nature of microbial growth. Physical factors such as fragmentation and embrittlement can also be assessed in these tests. Finally, the samples can be used to 'bait' microorganisms involved in the degradation process. These microbes, once isolated and characterised, can be incorporated into the petri dish screen as alternatives or additions to the current list.

The limitations in extrapolating the data from soil burial tests should be borne in mind. This is due to the necessary use of disturbed soil optimised in its characteristics for microbial growth and the higher temperature of incubation in relation to environmental conditions. The test, as with all those described in this chapter, should be regarded at best as an acceleration of environmental conditions and at worst as a relative indication of degradability which can be used to rank materials under the conditions of the test. However, the author's experience using plastics in soil burial tests is that a 3 to 6 month test is sufficient to demonstrate the environmental resistance of polymer materials.

6.9 Test method developments for the future

Industry and standards authorities have recognised the need to develop specific protocols to define and evaluate the biodegradable polymers.

Methodologies will be based upon those described in this chapter, but will need to be customised and fine tuned to reflect the definitions which are being debated at present.

ASTM has developed tests for evaluating degradable plastics in the aerobic and anaerobic sewage treatment environment. In 1991, CEN for Europe set up a standardisation committee (TC261) for setting standards and regulations for packaging and the environment. Working Group 2 of sub-committee 4 is concerned with degradability of packaging materials but has not yet produced any draft test methods. In Japan, a programme of work is under way using field exposure tests in soil, freshwater and the marine environment. Eighteen companies are involved in the trials and results are expected in 1993. The German DIN organisation has set up a committee (DIN 103.2) to address biodegradable polymers.

Environmental fate data have only been briefly mentioned in this description of test protocols. However, the need for specific data on the degradation pathway may assume increased importance. If this becomes the case, it will be necessary to develop test methods to follow metabolism of polymers. Such protocols are already in place for agricultural pesticides in soil, water and sediments and involve periodic sampling and extraction of metabolites from soils, water or sediments amended with the test substance. The substance is normally radiolabelled to assist in tracing and quantifying intermediate degradation products. Identification of significant intermediates will be necessary and an assessment of their ecotoxicity be required.

References

1. Fukuda, K. (1992) An overview of the activities of the Biodegradable Plastics Society. In *Biodegradable Polymers and Plastics* (eds M. Vert *et al.*), Royal Society of Chemistry, pp. 169–175.
2. Shuttleworth, W. A. and Seal, K. J. (1986) A rapid technique for evaluating the bideterioration potential of polyurethane elastomers. *Appl. Microbiol. Biotechnol.*, **23**, 407–409.
3. OECD (1981) *Guidelines for Testing Chemicals*, OECD, Paris.
4. Howard, P. H., Boethling, R. S., Jarvis, W. F., Melyan, W. M. and Michalenko, E. M. (1991) *Handbook of Environmental Degradation Rates*, Lewis Publishers, Michigan.
5. Muller, R. J., Augusta, J. and Pantke, M. (1992) An interlaboratory investigation into biodegradation of plastics. Part 1: A modified Sturm test. *Material und Organismen*, **27** (3), 179–189.
6. Seal, K. J. and Pantke, M. (1986) An interlaboratory investigation into the biodeterioration of plastics with specific reference to polyurethanes. Part 1: Petri dish test. *Material und Organismen*, **21** (2), 151–164.

7 Gelatinised starch products

G. J. L. GRIFFIN

7.1 History

Starch is the product of an industry based almost entirely on agricultural raw materials and has a world output measured in many millions of tonnes. Despite its chemical similarities with cellulose it has a spherulitic crystalline particle form very different from the familiar fibrous structure of vegetable cellulose and it is not usually regarded as a solid material in any engineering sense of the word. This is largely a consequence of its industrial applications which generally involve an initial stage of dissolution in water and it is regarded, therefore, as a thickening or binding agent and is mostly consumed by the paper, food, oilwell drilling mud, and textile trades apart from that part of the production which is converted hydrolytically into sugars. Its elastic modulus, however, as measured by ultrasonic pulse velocity measurement in starch particle/polymer composites [1], is very similar to pure cellulose and it is very surprising that more use has not been made of starch as a potential load bearing solid substance. The use of starch as a solid particle filler for plastics is discussed further in chapter 3 and it could be a technically superior alternative to woodflour in the making of heavily filled thermoplastic materials because it is colourless and does not generate organic acids pyrolytically, a phenomenon that has caused problems with the use of wood. There is, however, also a substantial literature of efforts to make thermoplastic materials by the esterification of starch (see, for example, Seiberlich [2]), very much in parallel with the successful production of acetates, propionates and mixed acetate/butyrate esters from cellulose. These cellulose esters provided the materials which enabled the processes of film casting, injection moulding and extrusion of thermoplastics to be established but, before the similar esters of starch could reach perfection, the field was taken over by the fully synthetic thermoplastics such as polystyrene, polymethyl methacrylate, and finally the polyolefines.

The early 1950s saw a significant investment of research effort into exploiting the fact that natural starch was a true high molecular weight polymer. It was soon appreciated that the amylose fraction was a preferable candidate for examination as a useful solid material, whereas the differences between amylose and amylopectin were not so important in the applications for starch in water solutions, the usual route into commercial applications of starch in the paper industry. Varieties of maize were identified which yielded a very high concentration of amylose in their starch, and success was reported [3] by Wolff *et al.* in converting this material into transparent films by casting techniques from water solutions, and by Carevic [4] whose patent reports the conversion of amylose into transparent films and tubes by extruding its dispersions in sodium hydroxide solution into acid coagulant baths. The technology never advanced beyond pilot plant demonstrations because of the established position of regenerated cellulose in the packaging market, and both cellulose films and amylose products were overtaken by the development of excellent extrusion blown films using low density polyethylene and, later, orientated polypropylene and high density polyethylene.

In the late 1960s the US Department of Agriculture initiated a programme of studies at the Northern Regional Development Laboratories in Peoria targeted at identifying and developing uses of cereal products for non-food applications. This concept had its roots in earlier fields of activity which had acquired the title of Chemurgy and one can cite the large scale commercial production of furfural from bran hulls as an instance of a successful outcome. Publications began to flow from the USDA labs on the subject of using starch oxidised by periodic acid to form a very reactive aldehyde which had promise of being a useful intermediate in rubber and polymer technology [5]. Later papers examined the possibility of blending starch with plasticised PVC as an extender/modifier and the conclusion was that coprecipitated starch gel with PVC compositions was a potentially useful material [6]. Unfortunately the technology of filled PVC systems using mineral fillers was so firmly established with excellent water resistance and sound economics that there was no commercial interest in these early starch/polymer blends. However, the team working on these materials, Westhoff, Otey, Doane, Mark and colleagues, pursued the study further to examine the possibility of blending gelatinised starch with polyolefines and other more polar synthetic polymers. The search for compatible systems eventually involved other centres such as the Battelle Institute and the laboratories of some of the major starch companies. Not surprisingly, water-soluble synthetics such as polyvinyl alcohol were examined as additives to starch gel formulations, as well as copolymers of vinyl alcohol with other monomers, and an astonishing harvest of patent filings has resulted.

7.2 Gelatinised starch blends with synthetic polymers

The fundamental problem was that the highly hydrophilic starch gel containing significant amounts of residual water was quite incompatible with, for example, polyethylene. A compromise was found by examining copolymers of ethylene with more polar monomers and the best candidate was ethylene/acrylic acid copolymer which was, fortunately, available as a commercial product. In order to achieve an adequately homogeneous melt these workers found that it was necessary to include a significant amount of concentrated ammonia solution in the formulation and, preferably, a small quantity of urea which has been noted to have a plasticising effect on gelatinised starch. Having achieved a composition which could actually be processed through a normal single screw extruder they experimented with additions of polyethylene and found a limited tolerance for this hydrophobic ingredient which, judging from freeze fracture microscopy, was present in the products as small frozen droplets. The existence of a product which could be made into thin films by the process of extrusion blowing justified public announcement of the work in 1974, much encouraged by the growth of interest in potentially biodegradable plastics. The work was described in published literature [7], and was also protected by filing patents from the US Government [8–10]. These patents, following the established policy of the time, were 'license of right' grantings available to any US citizen for exploitation provided that the citizen could convince the USDA that they were financially and technically capable. The licences were, however, non-exclusive and, probably as a consequence of this circumstance, there was no immediate commercial move to exploit the USDA invention and it slept in its file until the licensing rules were changed in about 1987. Efforts were then made by a small company in the USA, Agri-Tech Industries Inc., specially formed for the purpose. They took up an exclusive licence from the USDA and attempted to improve on the USDA technology and to organise manufacturing sub-licences with companies in the film extrusion business. Their target markets were principally the users of agricultural mulch film and garbage bags. This commercial adventure appeared not to prosper and, with the retirement of Otey from the USDA, even the work at the Peoria laboratories was apparently reduced in scale. In 1990 a version of the same technology was announced in Italy by the Ferruzzi Company, doubtless encouraged by the Italian environmental legislation which had imposed a levy on all polyethylene film bags which were not certified as biodegradable. The regulations were amended after about a year to impose a specified test for biodegradability, rather than relying on certification by university laboratories using the old ASTM fungal overgrowth test [11]. The ASTM test would give positive results with plastic films which had no more than surface contamination and the results were not, therefore, a true indication of biodegradability of the plastic films. The 'new' test demanded by the Italian authority was an adaptation of the old Sturm test originally applied to monitor water-soluble materials such as detergents. The starch based plastic films needed to be finely powdered to be acceptable in the test protocol and also to have a starch content of at least 40% to approach the degree of bacterial conversion of contained carbon into carbon dioxide demanded by the official specification. The Ferruzzi Company invested a great deal of money in the technical development of the manufacturing process, although to judge from their first patent filed [12] the formulations revealed were essentially identical with those featured in the USDA patent using starch gel blended with ethylene/acrylic acid copolymer, urea and aqueous ammonia. Later publications suggested that a wider variety of synthetic polymers was pressed into service, in particular the various grades and copolymers of polyvinyl alcohol.

Claims have been made that high shear mixing of these formulations, usually with twin screw compounding extruders, leads to the formulation of novel physical states akin to interpenetrating networks of starch gel and synthetic polymer at the near molecular level (see for example Bastioli *et al.* [13]), and this concept is presented on the basis of thin section TEM studies as well as by inference from the rheological behaviour of the melts and the changes that can be produced in the water sensitivity of the products by manipulating the mixing conditions during manufacture. Swanson *et al.* [20] and others at the USDA report that the presence of at least four phases can be proved in extrusion compounded blends of starch, polyethylene and ethylene acrylic acid copolymer using microscopy, nuclear magnetic resonance spectroscopy (NMR), X-ray diffraction and differential scanning calorimetry (DSC). Their conclusion was that the shear history of the blends determines the morphology of the final systems which, essentially, agrees with Bastioli and her colleagues. Of course, the presence of a continuous starch phase will ensure rapid access of microbial exo-enzymes to the starch content of the formulations but will, unfortunately, decrease the resistance of the materials to dramatic changes in physical properties on extended exposure to water. A new company, Novamont spa, was created to market the new products and their wares, launched in 1991, were prominently on display at the K92 exhibition in Dusseldorf under the trade name Mater-Bi. At about the same time the Italian government changed the regulations relating to plastics packaging by confirming the levy but eliminating the privileged position of degradable plastics. This action, not surprisingly, greatly depressed the market for any new materials. These starch matrix materials undoubtedly disintegrated quite rapidly in moist environments with rich microorganism populations and would lose about half of their weight into the surrounding humus. The fate of the synthetic polymer phase is less clear because of the high molecular weight of the copolymers used and we still await further developments before a fully biodegradable claim can be made. The film products appear to be adequately strong for general use provided only that there is no exposure to water which, because of the hydrophilic character of the continuous phase will result in swelling and loss of mechanical strength. Conversely, the films are quite resistant to oily materials and are agreeably free from problems of static charge build-up. The cost of the synthetic copolymers and the processing requirements of the manufacturing stage conspire to maintain the market price of these products at about three times the price of similar items made from LDPE. Economies of scale will probably reduce the price but it seems unlikely to drop below twice the price of LDPE. If the starch gel/polymer items could be collected after use there would seem to be no reason why they should not be recycled for re-use, subject only to the usual problems of recycling operations such as printing residues, labels, staples, bottle closures and the cost of collection. They could not, however, be processed in the current water washing systems used for recycling polyolefine materials and they would not be acceptable in admixture with polyolefine feedstocks.

7.2.1   Mater-Bi™ film extrusion

The Mater-Bi extrusion technology does not differ greatly from that of polyolefins and can use the same machines designed for PE tubular films.
   The grades recommended by Novamont for film extrusion are:

AF05H
AF10H

The peculiar rheology of Mater-Bi melts, can, however, require some modifications so as to ensure the best operating conditions. Most important, to get films of good mechanical strength and smooth surface, it is mandatory to extrude Mater-Bi at high shear rates and melt pressures. The rheological behaviour of the melts, approximating to Bingham characteristics as evidenced from Figure 7.1, signifies that high shear must be rapidly achieved and maintained. Mater-Bi is also heat sensitive, as indicated in Figure 7.2, and the dwelling time of the melt must be as short as possible in the extruder barrel avoiding any stagnation of the melt itself. The length/diameter ratio (L/D) of the chosen screw must be lower than 30. The most advisable L/D ratios are 25/30. Lower (20–25) ratios can result in incomplete plastification. Maillefer screws yield slightly lower outputs, but produce a better pigment dispersion when coloured master batches are used.
   The temperature control must be tight; the heat built up is higher than that with polyolefines due to higher friction with the barrel walls. The useful working temperature range is very narrow: 5 to 10°C. The extruder

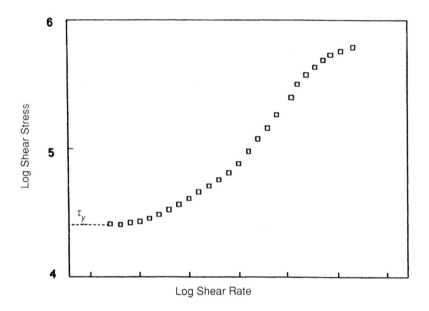

Figure 7.1 Melt flow behaviour of typical Mater-Bi material.

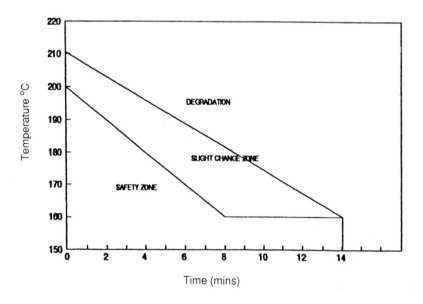

Figure 7.2 Thermal stability in processing of typical Mater-Bi materials.

motor must be of adequate power and able to overcome the initial high melt viscosity when starting the extrusion. As a reminder, compared with a grade 2 LDPE, Mater-Bi demands a 30% higher current drain at the same temperature settings and screw RPM.

The extruder screws should have a simple profile with a compression ratio (R/C) lower than 3. The best R/C values are between 2 and 3. The longer the screw, the higher should be the R/C. Dynamic mixing devices must be avoided. The best extruder screw types are:

(a) constant taper screws which, although rarely used, yield the maximum output rates;
(b) Maillefer screws which yield slightly lower outputs but produce a better pigment dispersion when coloured master batches are used;
(c) three-zone screws (feed – compression – metering), most widely used and recognised as being of almost universal application.

In order to avoid excessive pressure build up it is recommended to run the extruders with no screen pack other than the screen support grid if it has 0.3 to 0.5 mm holes. The dies should have a gap width of 0.5 to 0.7 mm for Mater-Bi™ grade AF05H and a gap of 0.5 mm for grade AF10H. For very large dies the latter gap can be increased to 0.6 mm. Typical extrusion settings are shown in Tables 7.1 to 7.3 for the three screw types mentioned. The consequences of gap variation are shown in Tables 7.4 and 7.5, and the outputs of the three screw types are compared in Table 7.6. It is normal practice with these materials to start up the extruders using LDPE alone, and, when the conditions have stabilised, introduce the desired grade of starch based compound. Conversely, before shut down the line should be purged with further LDPE. This ensures that the machines are not left to cool slowly with starch material cooking in the barrels and heads.

7.2.2   Mater-Bi™ injection moulding

Moulding grades need to display easier flow than extrusion materials and Novamont have formulated four grades to suit various moulding applications; these are listed in Table 7.7. The Bingham behaviour of these starch based compounds can give rise to mould filling problems unless it is appreciated that the critical minimum shear value is quickly reached and maintained until the cycle flow has finished. Table 7.8 lists the recommended moulding condition for the four grades offered. It will be noted that the working temperatures are lower than is usual for the common thermoplastics and it has to be remembered that the range of temperature tolerance is small, as was shown in Figure 7.2. Moulding machines with excellent temperature control systems are to be preferred; failure to fill cavities should be remedied by altering the gate and runner sizes and by adjusting the injection rate, never by simply raising the temperature. Hot

Table 7.1 Film extrusion conditions for Mater-Bi grade AF05H using a simple taper screw

Temperature profile (°C)		
Cylinder 1	140	
Cylinder 2	140	
Cylinder 3	140	
Cylinder 4	140	
Filter	140	
Neck	145	
Head	145	
Die	145	

	Melt temperature (°C)	Pressure (bar)
Cylinder	154	409
Neck	155	342
Die	151	175

Mechanical properties of film (at 23°C and 55% RH)	
Tensile strength (MPa)	20.4
Elongation at break (%)	409
Tensile modulus (MPa)	107
Energy at break (kJ m^{-2})	2609

Extruder ϕ = 40 mm; L/D = 30:1
Rotating spiral head: die ϕ = 100 mm
Screw: constant taper, R/C = 2.8; die gap 0.7 mm
Screw rotation: 64 rpm; blow-up ratio = 3:1
Output (kg h^{-1}): 32.05

Table 7.2 Film extrusion conditions for Mater-Bi grade AF10H using a Maillefer type screw

Temperature profile (°C)		
Feeding zone (grooved)	120	
Cylinder 1	135	
Cylinder 2	140	
Cylinder 3	140	
Cylinder 4	140	
Filter	140	
Neck	140	
Head	140	
Die	140	

	Melt temperature (°C)	Pressure (bar)
Neck	141	NA

Mechanical properties of film (at 23°C and 55% RH)	
Tensile strength (MPa)	22.2
Elongation at break (%)	393
Tensile modulus (MPa)	271
Energy at break (kJ m^{-2})	3161

Table 7.2—*continued*
Extruder ϕ = 55 mm; L/D = 30:1
Rotating spiral head: die ϕ = 150 mm
Screw: Maillefer: die gap 0.5 mm
Screw rotation: 65 rpm; blow-up ratio = 3:1
Output (kg h^{-1}): 65

Table 7.3 Film extrusion conditions for Mater-Bi grade AF05H using a metering type screw

Temperature profile (°C)	
Cylinder 1	140
Cylinder 2	140
Cylinder 3	140
Cylinder 4	140
Filter	145
Neck	145
Head	145
Die	145

	Melt temperature (°C)	Pressure (bar)
Cylinder	154	332
Neck	155	265
Die	147	145

Mechanical properties of film (at 23°C and 55% RH)	
Tensile strength (MPa)	16.3
Elongation at break (%)	451
Tensile modulus (MPa)	82
Energy at break (kJ m^{-2})	2049

Extruder ϕ = 40 mm; L/D = 30:1
Rotating spiral head: die ϕ = 100 mm
Screw: metering R/C = 3; die gap 0.7 mm
Screw rotation: 64 rpm; blow-up ratio = 3:1
Output (kg h^{-1}): 19.36

Table 7.4 Influence of die gap on film quality using Mater-Bi grade AF05H and metering type screw

	Die gap (mm)		
	0.5	0.7	0.9
Melt temperature (°C)			
Cylinder	157	154	156
Neck	156	155	155
Die	145	147	146
Melt pressure (bar)			
Cylinder	418	332	288
Neck	371	265	219
Die	242	147	96
Film appearance	Very good	Good	Rough, waved

Extruder ϕ = 40 mm; L/D = 30:1
Die ϕ = 100 mm; screw: metering
Temperature profile: 140 → 145°C
Screw rotation: 64 rpm; blow-up ratio = 3:1

Table 7.5 Influence of die gap on film quality using Mater-Bi grade AF10H and simple taper screw

	Die gap (mm)		
	0.5	0.7	0.9
Melt temperature (°C)			
Cylinder	156	158	158
Neck	149	148	149
Die	145	142	142
Melt pressure (bar)			
Cylinder	398	380	365
Neck	318	294	269
Die	158	106	91
Film appearance	Good	Slight white microwaves	Rough and pin holes

Extruder ϕ = 40 mm; L/D = 30:1
Die ϕ = 100 mm; screw: constant taper
Temperature profile: 135 → 140°C
Screw rotation: 64 rpm; blow-up ratio = 3:1

Table 7.6 Influence of screw type on output of film made from Mater-Bi grade AF05H

	Screw type		
	Constant taper	Maillefer	Metering
Compression ratio	2.8:1	2.4:1	3:1
Melt temperature (°C)			
Cylinder	152	155	154
Neck	155	155	155
Die	151	151	147
Melt pressure (bar)			
Cylinder	409	303	332
Neck	342	280	265
Die	175	140	145
Output (kg h^{-1})	32.05	25.32	19.36

Extruder ϕ = 40 mm; L/D = 30:1
Die ϕ = 100 mm; Die gap 0.7mm
Temperature profile: 140 → 145°C
Screw rotation: 64 rpm

runner systems have worked with starch based materials but excellent temperature control is essential. Possible ejection problems are minimised by using round or tapered runners and high finishes on mould surfaces; stainless steel is not necessary but polished chromium plate finishes are good. Unlike most injection moulding materials, which need to be kept dry, the starch gel products must maintain a water content within fairly close limits, therefore bags should be sealed to prevent water vapour loss in storage. Scrap mouldings and sprues can be granulated and reused if

Table 7.7 Physical properties of Mater-Bi injection moulding grades

Property	Unit	Test method	A105H	SA031	SA029*	Z101U
Melt flow rate, MFR (170°C, 5 kg)	g/10 min	ASTM D 1238	3	25	0.7	2
Fluidity: spiral path (170°C, 1500 kg)	mm	Internal met.	450	750	285	400
Tensile strength (1), (2)	MPa	ASTM D 638	20	14.5	29	>15
Elongation at break (1), (2)	%	ASTM D 638	100	615	5	>870
Tensile modulus (1), (2)	MPa	ASTM D 638	1300	250	4500	487

*20% glass fibre filled
(1) Dumbell injection moulded specimen, according to ASTM 2146
(2) Test conditions: 23°C, 55% RH

Table 7.8 Recommended moulding conditions for Mater-Bi injection moulding grades

	Unit	A105H	SA031	SA029	Z101U
Barrel temperature 1	°C	150	140	155	150
Barrel temperature 2	°C	160	155	165	165
Barrel temperature 3	°C	170	160	175	170
Nozzle temperature	°C	175	170	180	175
Injection speed	% max. speed	>70	>70	>70	>70
Injection pressure (melt)	bar	1200–1500	600–1000	>1200	1200–1500
Screw rotation	rpm	60–100	60–100	60–100	60–100
Mould temperature	°C	15–25	15–25	15–50	15–20
Shrinkage	%	0.6	0.6	0.6	0.6

quickly blended with about four times their weight of new material in order to compensate for loss of volatiles. Set-up and purging is best performed with LDPE, as for the extrusion operation. It is important to avoid raising the barrel temperature until all the starch based material has been cleared from the machine.

7.2.3 Properties of products

These compounds, with their high content of gelatinised starch, are, of course, quick to absorb moisture from the atmosphere and a typical plot of equilibrium moisture content against atmospheric relative humidity is shown in Figure 7.3. The consequences of taking up such a large amount of water, which acts as a plasticiser, are made clear in Figures 7.4 and 7.5. Conversely, of course, variations in storage or service temperature will alter the physical properties, and equilibrium values are shown in Figure 7.6. The time taken for films or mouldings to react to their environment will depend greatly on the dimensions of the parts and also upon the degree of circulation of the air around the parts.

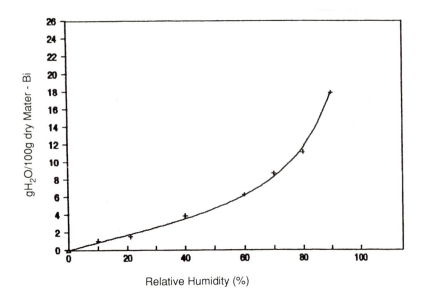

Figure 7.3 Equilibrium water absorption of Mater-Bi grade AF10H at 30°C.

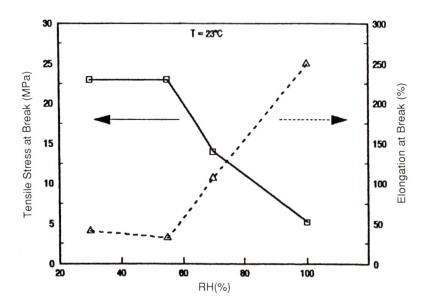

Figure 7.4 Physical properties of Mater-Bi grade AI05H as influenced by relative humidity.

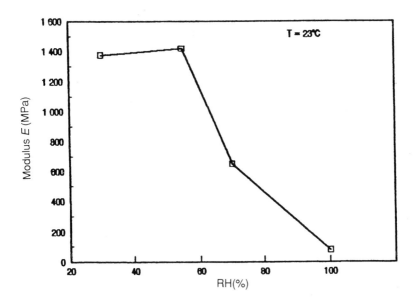

Figure 7.5 Variation in Young's modulus of Mater-Bi grade AI05H as influenced by relative humidity.

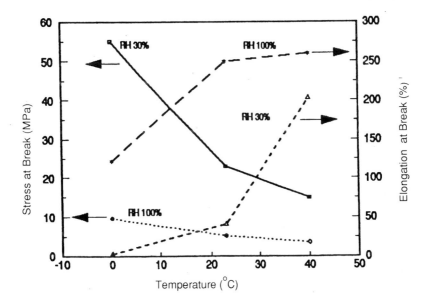

Figure 7.6 Physical properties of Mater-Bi grade AI05H as influenced by temperature.

7.3 High starch content products

Following the early work on amylose films, activity in the period up to the 1980s was minimal, and certainly not generally disclosed, with most public attention being focused on the starch/polymer blend materials and on photodegradable products. In 1987, however, Swiss and German newspapers reported the work of Tomka and Wittwer at the Eidgenossiche Technische Höchschule in Zurich describing a process for moulding starch, without added synthetics, as if it were a thermoplastic material. Detail eventually emerged in a patent published in the USA in 1987 [14] revealing that the work had reached a point at which patenting was considered worth while in early 1983. The tangle of early applications, some continued and some abandoned, is scarcely worth citing here and they can readily be traced by reference to the later published specifications.

Since any attempt to melt starch by simply heating it in an open vessel leads to dramatic pyrolysis without fusion, even in an inert atmosphere, it is clear that another substance must be present which could attach itself to the starch molecules and reduce the hydrogen bonding between the chains to a level at which true fusion could occur below the decomposition temperature. The obvious 'plasticiser' is water and Tomka and Wittwer evidently appreciated that the fusion process would have to take place at temperatures above the boiling point of water. Following the logical engineering step of running the process at high pressure they were successful in melting starch containing between 5 and 30% of water in a modified injection moulding machine. The patent document is dominated by a discussion of the type of injection moulding machine used which was, nevertheless, a perfectly normal modern machine familiar to the industry. The discussion in the patent about the behaviour of the starch is a precise description of the type of phase diagram applicable to the starch/water system at temperatures and pressures not available in the usual processes of starch dissolution in hot water. The melting process of the starch is attributed to the collapse of the helical crystalline structure associated with amylose, and the description is supported by display of a DSC record. Unfortunately, the DSC instrument cannot operate other than at atmospheric pressure and we are left to speculate on what might be happening within the injection moulding machines. The technology was encouraged by the existence of a market for medical capsules not made from animal protein, as is the case with the established traditional capsules cast from gelatine, and sponsorship came from the Capsugel Company, part of the Warner Lambert Co. This work was the subject of intensive development aimed at perfecting the moulding operation and identifying minor additives which could improve the processing ability of the materials and the mechanical properties of the products. It was also hoped to introduce a degree of hydrophobicity which would increase the tolerance of the mouldings to contact with water or humid environments. Several of these patents are listed in the references to this chapter [14–17] and it is certain that others have also been filed. The commercialisation is under the name Novon™. The original Tomka pure starch/water mouldings could legitimately claim to be fully biodegradable but this claim will need re-examination as other additives are explored. Once again the problem of strength loss in contact with water or water-containing materials somewhat restricts the range of applications and the question of cost seems to run parallel with the Novamont products. It has been proposed, by Stepto and Dobier [16], that the effect of shear and high temperatures on the starch in the Novamont technology is to change the starch into a mechanically superior form devoid of any crystallinity, as evidenced by X-ray diffraction and differential scanning calorimetry studies. The developers wish to identify this by the term 'destructured' starch. This claim has been strongly contested by the USDA team [19, 20] who insist that the terms gelatinised, destructured, cooked, etc. are all equal in significance and simply imply that the crystal structure of the starch spherulites is destroyed by gelatinisation at sufficiently elevated temperatures in the presence of a sufficient quantity of water. Novon, however, have been granted patents on the basis of destructured starch being accepted as a recognisable form of matter on the basis of its measurable physical properties. There are so many alternative formulations described in the Warner Lambert/Novon patents that it is rather difficult to quote significant physical data on the emerging materials. It is interesting, however, to remember that the original objective was to manufacture moulded solid objects from starch and water alone, and the older literature does offer us a guide to attainable strengths in similar systems made by casting from water solutions rather than by high pressure moulding. The data given in the Wolff *et al.* paper [3] from 1951, when converted into modern units, as in Table 7.9, is very interesting in that the

Table 7.9 Data from Wolff *et al.* [3] on starch films cast from water solutions and conditioned at 30% RH and 22°C

Amylose (%)	Tensile strength (MPa)	Elongation at break (%)
95	6.96	23
86	5.39	8
77	5.88	11
68	6.08	18
59	6.28	10
50	6.08	6
42	3.92	6
29	5.00	6
24	3.24	3
15	3.43	9
3	2.94	5

benefits of high amylose content are very clear, and also the tensile strengths attained seem almost appropriate to polyethylene.

Another Novon™ product has appeared in the form of an extruded starch foam, achieved by careful manipulation of the water content and die temperatures in the extrusion process and using the flash generation of steam as the starch melt leaves the die to achieve the desired expansion to a fine cell structure. The principle application target is a biodegradable alternative to the familiar polystyrene foam packaging pellets. A similar product is disclosed in a patent published in 1991 [18] by the National Starch & Chemical Investment Corp'n.

References

1. Griffin, G. J. L. (1985) In *New approaches to Research on Cereal Carbohydrates* (Eds R. H. Hill and L. Munck), Elsevier, pp. 201–210.
2. Seiberlich, J. (1942) *Modern Plastics*, **18**(7), 64–65.
3. Wolff, I. A., Davis, H. A., Cluskey, J. E., Gundrum, L. J. and Rist, C. E. (1951) Preparation of films from amylose. *Ind. Eng. Chem.*, **43**, 915–919.
4. Carevic (1967) US Patent 3,336,429.
5. Pfeifer, V. F., Sohns, V. E., Conway, H. F., Lancaster, E. B., Dabic, S. and Griffin, E. L. (1960) *Ind. Eng. Chem.*, **52**, 201–206.
6. Westhoff, R. P., Otey, F. H., Mehtretter, C. L. and Russell, C. R. (1974) Starch filled PVC plastic—preparation and evaluation. *Ind. Eng. Chem. Prod. Dev.*, **13**(2), 23–25.
7. Otey, F. H., Westhoff, R. P. and Doane, W. M. (1980) Starch based blown films. *Ind. Eng. Chem. Prod. Res. Dev.*, **19**(4), 592–595.
8. Otey, F. H. and Mark, A. M. (1976) Degradable starch based agricultural mulch film. US Patent 3,949,145.
9. Otey, F. H. and Westhoff, R. P. (1979) Biodegradable films from starch and ethylene acrylic acid copolymers. US Patent 4,133,784.
10. Otey, F. H. (1982) Biodegradable starch based blown films. US Patent 4,337,181.
11. *Standard Practice for Determining the Resistance of Synthetic Materials to Fungi.* ASTM Procedure G21–70 (Re-approved 1985).
12. Bastioli, C., Bellotti, V., Del Giudice, L., Del Tredici, G., Lombi, R. and Rallis, A. (1990) PCT WO 90/10671.
13. Bastioli, C., Bellotti, V. and Gilli, G. (1990) Agricultural commodities as a source of new plastic materials. Paper presented at *Biodegradable Packaging and Agricultural Films*, Paris, Mary 1990.
14. Wittwer, F. and Tomka, I. (1987) US Patent 4,673,438.
15. Lay, G. J., Rehm, R. F., Stepto, R., Thom, M., Lentz, D. J., Sachetto, J. P. and Silbiger, J. (1992) US Patent 5,095,054.
16. Stepto, R. and Dobier, B (1989) EP Application 0326517.
17. Sachetto, J. P., Silbiger, J. and Lentz, D. J. (1990) EP 409,781 A2.
18. Lacourse, N. L. and Altieri, P. A. (1991) Biodegradable shaped products and method of preparation. US Patent 5,035,930.
19. Shogren, R. L., Fanta, G. F. and Doane, W. M. (1993) *Die Stärke*, **45**(8), 276–280.
20. Swanson, C. L., Shogren, R. L., Fanta, G. F. and Imam, S. H. (1993) *J. Environ. Polym. Degrad*, **1**(2), 155–166.

Index

aerobic organisms 10
Alcaligenes eutrophus 50
aliphatic polyanhydrides 16
amylose cast films 149
anaerobic conditions 12
anaerobic sewage digester 121
annealing behaviour, starch LDPE films
 41
antioxidants 39
Archer Daniels Midland Co. 23
Arrhenius equation 102
autoxidation 24
    measurement by incubation 40
autoxidants, balance with antioxidants 39

Bacillus cereus 15
Bacillus megaterium 80
back pressure, in PE/starch extrusion 102
biodegradability, defined in standards
    ASTM D20.96 proposal 117
    DIN 103.2 117
    ISO 472:1988 117
    Japanese Biodegradable Plastics Society,
        draft 117
biodegradability evaluation 116
biodegradation
    accelerated testing 39
    evaluation criteria 118
    municipal composting facilities 46
    PHBV *in vivo* 74, 77
    political expediency 116
    polycaprolactone 5
    polyethylene 13, 14
    poly (hydroxyalkanoates) 68–71, 73–75
    proof using radioisotopes 45
    soil burial tests 45
    Sturm test 46
    tested in activated sewage sludge 46
    testing strategies 119
    undesirable at 100% level 117
biodegradation testing
    application simulation 129
    carbon isotope labelling 127

environmental chamber tests 131–132
future developments 133
reproducibility 125
soil burial tests 132–133
soil types for burial 128, 132
biological hydrolysis 11
biological oxidation 9
Biopol™ 50
biopolyesters 48
blocking 24, 35
blow up ratio (BUR) of PE/starch film 109
bottle blowing, starch HDPE
    compounds 33
bottles, Biopol™ 50
Brunel University 5, 18, 23
Buss KoKneader 27

calcium oxide, desiccant 101, 105
calcium oxide masterbatch 105
carbon dioxide evolution monitoring
    123–129
carbonyl index versus time and
    temperature 42
carrier resins for pigments 98, 99
Carson, Rachel 1
Chromobacterium violaceum 55
closed bottle test 126
coenzymes 8
cofactors 8
Coloroll Ltd 18
compostable bags, desirable properties 40
composting 3, 4
    accelerated 20
composting miniaturisation problems 121

Dano system, composting 21
decision trees, test methodology
    selection 121
degradation acceleration after recycling
    105
degradative pathway, complete 118
desiccant, calcium oxide 101, 105
deterioration on reprocessing 98

dextran gum, surface film   44
die face cutter   104
diene rubbers   24
dies, extrusion, for PE/starch films   100
differential thermal analysis   19
disposable items   2
disposal route simulation   120, 123
dissolved oxygen concentration (DOC)
　123
DTA *see* differential thermal analysis   19

Ecostar Ltd.   21
Ecological Materials Research Institute   23
energy to break, starch LDPE films   35
environmental chamber biodegradability
　tests   131, 132
enzymatic degradation   7
enzymes
　cytochromoxidase   10
　depolymerase   51
　hydroxylases   10
　nomenclature   10
　proteolytic   11
　specificity   10
EPRON Industries Ltd.   23
extrusion coating   31

face cutting   30
fat, rancid   21
fatty acid esters   23
Ferruzzi Ltd.   137, 138
fungal overgrowth tests in Petri dishes
　130, 137

gelatinised starch
　compatibility with copoly(ethylene
　　acrylic acid)   137, 138
　compatibility with polyvinyl alcohol
　　136, 137
　foam extrusion   150
　fusion under pressure   148
　injection moulding   148
　phase diagram with water   148
　water plasticisation   148
　X-ray diffraction studies   149
gelatinised starch/polymer blends
　Bingham body behaviour   141
　film extrusion   139, 141
　friction in extrusion   139
　injection moulding   141
　physical properties   145
　price   139
　recycling possibility   139
　rheology   139
　water sensitivity   139
gene mutation   13
genetic engineering, PHB genes in *E. coli*
　and plants   87

half life versus temperature for starch
　LDPE films   41
hydrolysis
　of cellulose acetate   16
　of polyhydroxybutyrate   16, 72
　of pullulan   16
hydrophobic surfaces   15
hydrophobic treatment   20
hydroxyalkanoic acids   48
hydroxybutyrate valerate copolymer   50,
　52, 54
hyphal penetration   12

incineration   3
induction period
　by thermogravimetric analysis   42
　log plot versus temperature   44
　measurement   41
　thermal   37, 41
inherently biodegradable materials   120,
　126
injection moulding, starch polyolefin
　compounds   33
interpenetrating networks   138
Italian Government levy on plastics
　packs   138

landfill   3
landfill technology   97
Lemoigne, Maurice   49
lignin   7
long side chain (LSC)
　poly(hydroxyalkanoates)   80–84

macroorganisms   7
market potential, USA, for degradable
　films   99
marine environments   121
masterbatch, starch LDPE   20
Mater-Bi™   138
melt flow index   31
metabolisation by microorganisms   44
MFI *see* melt flow index   31
microbial sludge   127
mildew   12
mineralisation proof by $^{14}CO_2$ evolution
　127
moisture absorption, PE/starch materials
　30, 102
moisture pick up   30
molecular oxidation   11
molecular size
　effect of autoxidation   44
　measurement by GPC   42
　reduction by oxidation   39
morphology of gelatinised starch/polymer
　blends   138
MSW *see* municipal solid waste   2

mulch film 137
municipal solid waste 2, 21

Nelson packaging Ltd. 21
Novon™ 149

OECD biodegradability test protocols 118
oxidation 7
  time controlled 39

particulate fillers 16
pelletising extruder 103
permeability of starch LDPE films 35
permeability to proteins 20
peroxide formation 21
PHB see poly(hydroxybutyrate) 48, 49
PHB and PHBV blends with
  cellulose esters 87
  polyethylene oxide 86
  polyvinyl acetate 86
  polyvinyl chloride 86
  polyvinylidine fluoride 86
  starch 86
PHBV cost, factors influencing 78–79
photodegradation 2
photooxidation 13
physical properties, starch LDPE
  compounds 34, 36
plastic paper 16
Polyclean™ 23
poly(hydroxybutyrate)
  applications 75–78
  biodegradation like
    poly(caprolactone) 75
  bone fracture fixation 77
  brittle character 58
  complexes with lipoproteins 51
  crystal structure 61–65
  crystallisation kinetics 65–67
  enzymatic digestion 57
  extracellular biodegradation 71
  gas permeability 76
  hydrolytic degradation 72–73
  hypochlorite digestion 56
  in bacterial membranes 51
  inclusion bodies 49, 51
  intracellular biodegradation 68–71
  isodimorphism of copolymers 63
  isolation 55–57
  isomorphism of comonomers, NMR
    studies 62
  membranes around inclusion bodies 65
  metabolic synthesis 52
  molecular weight 52
  molecules per cell 52
  nascent morphology 65
  percent in bacterial cells 51
  physical properties 60

piezoelectric properties 77
plastic laminate 49
production quantity 49
recycling, possible problems 75
shelf life 74
shish kebabs 67
solubility related to crystallinity 67
solvent extraction 55–56
thermal degradation 68, 72
transition temperature 58
X-ray diffraction studies 61
poly(adipic anhydride), new synthesis 16
polycaprolactone 5
poly(ethylene succinate)-β-poly(ethylene
  glycol) 16
poly(ethylene succinate)-β-(tetramethylene
  glycol) 16
poly(β-hydroxyvalerate) 52–54
poly(β-propiolactone) 16
poly(tetramethylene adipate) 16
poly(trimethylene carbonate) 16
primary biodegradation of polymer
  molecule 118
processing equipment for PE/starch film
  99, 100
pro-oxidants in PE/starch compounds 102
property losses, biodegradation assessment
  by 131
Pruteen animal feed 50
Pseudomonas cepacia 55
Pseudomonas oleovorans 80
public acceptance, biodegradable plastics
  97
pyrolysis temperature of starch 31

radiolabelled test substances 127
readily biodegradable materials 120
recycling 25
  costs 103
  of PE/starch, cross contamination 111
  PE/starch film 97
  PE/starch film properties 108
  PE/starch material economics 107
reprocessing, film scrap 98
respiration data 119
rheology, starch/polyolefin compounds 31
risk assessment for plastics 116

saprophytic decay 12
scavenger, moisture 101, 104
screening tests for ready
  biodegradability 123
screening tests, short term, for
  biodegradability 120
sewage environments 121
shelf life from thermal ageing data
  114–115

simulation tests, aerobic sewage
  treatment 129
soil burial tests, sieving and moisture
  content 129
solvent extraction of PHB 49
starches
  agglomeration of particles 100
  amylose/amylopectin ratio 136, 149
  blends with PVC 136
  calorific value 25
  density 25
  dispersion in polymer melts 27
  elastic modulus 135
  endothermic pyrolysis 19
  energy consumption, manufacture 25
  esters 135
  gelatinised 136
  moisture content of 19, 101
  oxidised by periodic acid 136
  particle size distribution 25
  refractive index 26
  small particle size 24, 27
  thermal stability 35, 101
  voids in particles 26
St. Lawrence Starch Co. 21
Staudinger, J. J. P. 2
steric hindrance 14, 15
storage, reprocessed PE/starch
  compounds 106
storage life, masterbatch 30
Sturm test, modified 123, 124
sulphate reducing bacteria 12
surface attack by microorganisms on PHB
  and PHBV 73–75

synthetic poly(hydroxyalkanoates) 80–86

test sample making, starch LDPE
  compounds 34
testing environments for biodegradable
  plastics 120
TGA see thermogravimetric analysis 42
theoretical oxygen demand (TOC) 126
thermogravimetric analysis, isothermal 42
thickness of starch LDPE films by
  average, variation with gauge 101
  gravimetric measurement 35, 100
  micrometer measurement 100
tiered approach to evaluating
  degradability 119
tool design, injection moulding Mater-
  Bi™ 144
toxicity 23
transition metal catalysts 23
transverse elongation, effect of blow up
  ratio 110
twin screw compounding machines 30

United States Department of Agriculture
  (USDA) 4, 136

vertical integration, masterbatch making
  24
volatiles, loss of from starch LDPE film 42

Warner Lambert Co. 149
waste plastic, in-plant 98